P9-DNV-024

The Organic Chemistry of Peptides

The Organic Chemistry of Peptides

HARRY D. LAW
Head of Chemistry,
Liverpool Polytechnic

WILEY — INTERSCIENCE
a division of John Wiley & Sons Ltd
LONDON NEW YORK SYDNEY TORONTO

Library of Congress Catalog Card No. 74–93558

ISBN 0 471 51899 9

Made and Printed in Great Britain by William Clowes and Sons Limited, London, Beccles and Colchester

Preface

Peptide chemistry has made exciting advances over the last two decades and merging, as it does, into molecular biology, it will certainly continue to be the subject of intensive research efforts. On these grounds alone, peptides merit attention in undergraduate courses in organic chemistry, but their intrinsic importance is not the only justification for including them; I believe that an introductory treatment of peptide chemistry can also have considerable didactic value, a contention which I hope this book will support.

In this volume, not only have I tried to give an accurate view of modern peptide chemistry, but I have also endeavoured to use the subject matter to illuminate fundamental chemical concepts and principles. This has led me to a somewhat unusual construction in which a series of exercises forms an integral part of the text. These exercises, which are asterisked, should be attempted by the student before he proceeds to the next part of the text where, either directly or indirectly, they are discussed. I have adopted this approach, which has obvious affinities with programmed learning, so that I might put the student into research, or simulated research situations as they arise and thus lead him on to the next part of the theme. Other exercises, not asterisked, are somewhat aside from the main theme and the answers to these exercises are incorporated into an appendix.

This construction lends itself well to the first three quarters of the book which deal with the basic organic chemistry of peptides, but it is not so readily applicable to the final two chapters which build a bridge into molecular biology. Of necessity, these are more descriptive.

Regretfully, I have not been able to quote original references in the text. They are so numerous that their inclusion would have substantially increased the size of the book, a development which I thought unjustified, since most readers will not be concerned with them. A review bibliography is included for readers who want to go more deeply into the subject. I have also appended a chronological list of significant advances in the field, which may help to atone for the fact that my approach to peptide chemistry is hardly likely to give the student a sense of its historical evolution.

A student might be expected to cover the groundwork on which this book rests during the first two years of an Honours Degree course. In most cases, therefore, he will be ready to tackle peptide chemistry towards the end of the penultimate year or early in the final year of such a course. Candidates for Graduate Membership of the Royal Institute of Chemistry will find that the book consolidates much of the material which they covered at Part I level; it might profitably be included in the Part II course or provide suitable preparatory material for a final full-time year. These are the readers which I had in mind when writing the book, but certain parts of it will undoubtedly prove suitable for lower level courses, for example Higher National Certificate and Higher National Diploma Courses, whilst even research students working on peptide topics have found other parts instructive.

It is a pleasure to acknowledge the invaluable assistance of Professor G. W. Kenner, F.R.S. and Dr. P. G. Jones who read the entire manuscript as peptide-expert and non-peptide chemist respectively. They both made many improving suggestions without which the book would have been substantially the poorer. Of course, I am solely responsible for any errors or imbalance of emphasis which remain.

HARRY D. LAW

Little Sutton, June 1970

Contents

CHAPTER ONE

Significance and General Structure

Occurrence, composition and properties

Peptides are composed of α-amino acids joined together by amide bonds. The simplest of these compounds (**1**; $R^1 = R^2 = H$) possesses only two amino acid units per molecule, whereas more complex examples incorporate a large number of amino acids and, in terms of molecular weight, can be as much as ten thousand times larger. Peptides of relatively low molecular weight ($M <$ about 10,000) are called polypeptides, the higher molecular weight materials, proteins. Small polypeptides, sometimes referred to as oligopeptides, are given names which indicate the number of amino-acid units in the molecule; *dipeptides* (**1**) possess two such units; *tripeptides* (**2**) three; *tetrapeptides* (**3**) four; and so on.

$$NH_2.\underset{|}{\overset{R^1}{C}}H.CO.NH.\overset{R^2}{C}H.CO_2H$$

(1)

$$NH_2.\overset{R^1}{C}H.CO.NH.\overset{R^2}{C}H.CO.NH.\overset{R^3}{C}H.CO_2H$$

(2)

$$NH_2.\overset{R^1}{C}H.CO.NH.\overset{R^2}{C}H.CO.NH.\overset{R^3}{C}H.CO.NH.\overset{R^4}{C}H.CO_2H$$

(3)

The biological importance of peptides cannot be overstated. They occur in all living tissues and are fundamental to life itself. The enzymes, on which nearly all biochemical processes depend and through which the genes exercise their control over metabolism, are soluble proteins; as, for example, are haemoglobin, the oxygen-transporting pigment of blood; γ-globulin, from which antibodies are formed; and insulin, the hormone involved in the regulation of blood sugar levels. Some proteins are relatively insoluble and fulfil structural roles. Such proteins account almost entirely for hair, nails, horn, wool, silk, cartilage and muscle. Peptides of lower molecular weight are equally ubiquitous and vital. They include many hormones and innumerable metabolites of microorganisms which often possess antibiotic properties.

At first, it may seem surprising that compounds which are based upon such a simple structural unit can fulfil so wide a diversity of functions. However, the formulation of peptides as polyamides of α-amino acids is substantiated by a large body of evidence, which cannot be concisely stated, but which involves the accumulated results of physical, degradative and synthetic investigations. Similarly, it might be thought that the organic chemistry of peptides will be somewhat limited and uninformative, but this is not the case. The apparent simplicity of the system is deceptive.

It is true that peptide chemistry is relatively specialized, but this is mainly because of the peculiar technical difficulties involved. These difficulties aside, investigations of peptide structure and function call for a sound appreciation of the basic principles of organic chemistry, and studies of peptides may be used with singular success to illustrate the importance of these principles. Particular emphasis will be given to these considerations throughout this book; the practical difficulties, together with the techniques which have been developed to overcome them, will generally be disregarded. However, before dismissing them entirely, it is worthwhile to consider the nature of these difficulties.

*Exercise 1.1**

Ignoring stereochemical considerations, do you think that formulae **1**, **2** and **3** are likely to be accurate representations of the respective peptide molecules as they would exist in aqueous solution?

Although the di-, tri- and tetrapeptides (**1**, **2**, **3**) have been written for simplicity as un-ionized structures, peptides which possess free amino and free carboxyl groups will tend, like the amino acids themselves, to form dipolar ions or zwitterions (e.g. **4**). As a result, they often possess high, indefinite melting points and are soluble in aqueous rather than organic solvents. This leads to difficulties in characterization and purification.

$$\overset{+}{N}H_3.\underset{}{\overset{R^1}{\overset{|}{C}}}H.CO.NH.\overset{R^2}{\overset{|}{C}}H.CO_2^-$$

(4)

In addition, peptides frequently occur in complex mixtures with other peptides and have a facility for forming mixed crystals, so that even crystallinity is no indication of purity. With large peptides, there is the further consideration that the three-dimensional structure is both complex and important. The forces which determine this structure (Chapter 7) are easily disturbed with the result that the peptide loses its characteristic properties, often irreversibly. This process is referred to as *denaturation*.

The fundamental polyamide structure of peptides was understood at the turn of the century, but these various technical problems were so intractable that nearly fifty years elapsed before a meaningful approach could be made to the more detailed structural and hence functional chemistry of peptides. Chromatography in its various guises was the tool which permitted this further advance to be made and peptide chemistry still centres on chromatographic techniques. As a corollary, it is true to say that the technical difficulties of peptide chemistry provided perhaps the main stimulus in the early development of chromatography.

The hydrolysis of peptides

---

*Exercise 1.2**

By analogy with ester hydrolysis, postulate mechanisms for the hydrolysis of benzamide. Under alkaline conditions the hydrolysis of ^{18}O-benzamide proceeds with isotopic exchange; no exchange occurs under acidic conditions.

One compelling piece of evidence in favour of the polyamide theory of peptide structure is the observation that peptides (**5**) can be readily hydrolysed to give α-amino acids which usually account quantitatively for the weight of peptide treated.

$$\overset{+}{N}H_3.\underset{}{\overset{R^1}{CH}}.CO.NH.\overset{R^2}{CH}.CO.NH.\overset{R^3}{CH}.CO\ldots NH.\overset{R^n}{CH}.CO_2^- \xrightarrow{+(n-1)H_2O}$$

(**5**)

$$\overset{+}{N}H_3.\overset{R^1}{CH}.CO_2^- + \overset{+}{N}H_3.\overset{R^2}{CH}.CO_2^- + \overset{+}{N}H_3.\overset{R^3}{CH}.CO_2^- + \ldots NH_3.\overset{R^n}{CH}.CO_2^-$$

The hydrolysis of simple amides (**6**) has much in common with ester hydrolysis. Under alkaline conditions, nucleophilic addition of hydroxyl ion to the amide carbonyl occurs, followed by the elimination of ammonia ($R^1 = H$) or an amine.

$$R{-}\overset{O}{\overset{\|}{C}}{-}NHR^1 \underset{}{\overset{OH^-}{\rightleftharpoons}} R{-}\overset{O^-}{\underset{OH}{C}}{-}NHR^1 \rightleftharpoons R{-}\overset{OH}{\underset{O^-}{C}}{-}NHR^1$$

(**6**)

$$\downarrow$$

$$R.CO_2^- + NH_2R^1$$

Alkaline hydrolysis is only infrequently used in peptide studies and it will not be considered further at this stage. On the other hand, acidic hydrolysis is very important.

Under acidic conditions, protonation of the amide presumably precedes addition of water.

$$R{-}\overset{O}{\overset{\|}{C}}{-}NHR^1 \overset{H^+}{\rightleftharpoons} R{-}\overset{^+OH}{\overset{\|}{C}}{-}NHR^1 \rightleftharpoons R{-}\overset{O}{\overset{\|}{C}}{-}\overset{+}{N}H_2R^1$$

(**6**)

$$\overset{H_2O}{\rightleftharpoons}$$

$$R{-}\overset{OH}{\underset{^+OH_2}{C}}{-}NHR^1 \rightleftharpoons R{-}\overset{OH}{\underset{OH}{C}}{-}\overset{+}{N}H_2R^1 \rightleftharpoons R{-}C\begin{matrix}{=}O\\ {-}OH\end{matrix} + \overset{+}{N}H_3R^1$$

For simple amides, acidic hydrolysis, like most reactions which involve addition of a nucleophile to a carbonyl derivative, has an optimum acidity.

This is generally explained by saying that the amide is perhaps incompletely protonated in weakly acidic solutions, whereas, in more acidic solutions, the availability of the nucleophile (H_2O) is probably reduced.

These conclusions should only be extended with caution to the hydrolysis of peptides. The rate of hydrolysis of the protein, gelatin, for example, increases linearly with acid concentration over the range 3–10·4M (HCl). A much simpler model, 2,5-dioxopiperazine (7) behaves similarly. Most peptides are hydrolysed completely by treating them with constant boiling hydrochloric acid (6·5M) at 100–110°C for sixteen to twenty hours.

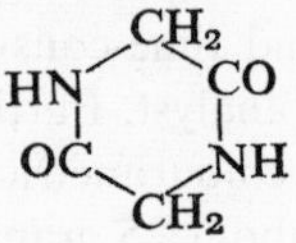

CH_2
HN CO
OC NH
CH_2

(7)

*Exercise 1.3**

The empirical formula of the protein ribonuclease (obtained from cattle) is:

$$C_{566} H_{890} O_{168} N_{192} S_{13}$$

It is obviously meaningless to use such a formula in the classical way as a basis for structural investigations. Can you suggest a more practical way of measuring the composition of a protein?

Amino acid analysis

When the complexity of the larger peptides is taken into account it is really rather fortunate that peptides can be so readily converted into amino acids. The elemental analysis of such large molecules, particularly when molecular weight uncertainties prevail, is not helpful. For example, in Exercise 1.3*, only the sulphur analysis would provide a useful check for degradative studies. On the other hand, a quantitative estimation of the amino acids produced by the hydrolysis of a protein, the 'amino acid analysis,' is invaluable. The amino acid residues (.NH.CHR.CO.) must be present in integral amounts so that, if the molecular weight of the peptide is known, the amino acid analysis will indicate how many residues of each amino acid are present in the molecule; if the molecular weight is unknown, the amino acid analysis will permit the minimum possible molecular weight to be calculated and will reveal the relative proportions of the different amino acid residues.

Although the range of possible α-amino acids is infinite, luckily only twenty or so are found as constituents of proteins. These are referred to as the common α-amino acids (Table 1.1). Even so, until the development of chromatographic techniques for separating amino acids, amino acid analyses were very laborious and time-consuming and called for a high order of skill on the part of the analyst. Furthermore, the analyses, which were based on gravimetric and colourimetric procedures, consumed considerable amounts of peptide (about 25 grams of an average protein was required for a full amino acid analysis). By contrast, the best of the chromatographic procedures can be carried out by a semiskilled operator, requires no more than 0·1 μmole of protein and gives a complete amino acid analysis in less than four hours. In this procedure, the amino acid mixture at pH 2·2 is applied to the top of a column of resin which consists of partially sulphonated polystyrene beads. At pH 2·2, the amino acids act as cations and are ionically bound by the sulphonic acid groups. The column is developed with aqueous buffer solutions of increasing pH and the amino acids are thereby eluted from the column. Partition, van der Waal's adsorption and other factors, in addition to the base strengths of the amino acids, contribute to determine the ease with which individual amino acids are displaced from the resin. A pattern of elution has been found empirically which brings about the separation of all of the common amino acids. These are detected and estimated colourimetrically in the effluent by the colour produced when amino acids react with ninhydrin (**8**). The point of emergence of an amino acid in the effluent is characteristic of the amino acid.

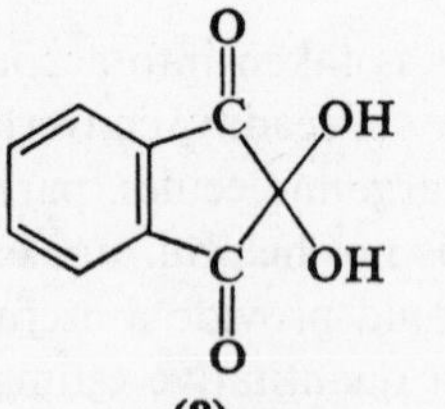

(8) 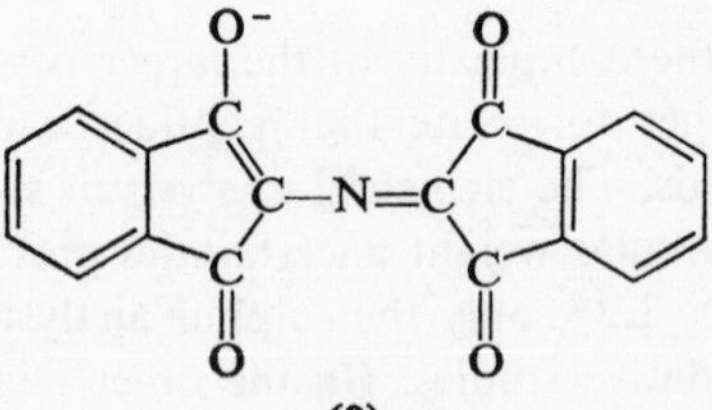 **(9)**

Exercise 1.4

Gem-diols (hydroxyl groups attached to the same carbon atom) are usually expected to be unstable. What factors are responsible for the stability of ninhydrin in the diol form? (Compare with chloral hydrate.)

Exercise 1.5

α-Amino acids react with ninhydrin to give a blue-coloured product, diketohydrindylidene-diketohydrindamine (**9**). The reaction sequence involves Schiff base formation, decarboxylation, imine hydrolysis and Schiff base formation. Outline this series of changes.

It is often difficult to obtain amino acid analyses for peptide metabolites of microorganisms, because amino acids are encountered, for example ornithine, $NH_2CH(CH_2CH_2CH_2NH_2)CO_2H$, which are not found as constituents of proteins. In addition, whereas protein amino acids are invariably of the L-configuration, those derived from microorganisms can be L or D. For these reasons, it is often necessary when investigating these peptides to isolate and characterize the amino acids involved.

Exercise 1.6

L(−)Serine has the structure shown in the Fischer projection formula (**10**). It is therefore of the *S*-configuration. How is the configuration of L(+)alanine (**11**) and L(−)cystine (**12**) described in these terms? How many stereoisomeric forms of cystine are there?

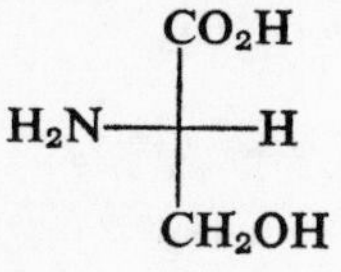

$$\begin{array}{c} CO_2H \\ H_2N\text{—}\!\!\!+\!\!\!\text{—}H \\ CH_2OH \end{array} \qquad \begin{array}{c} CO_2H \\ H_2N\text{—}\!\!\!+\!\!\!\text{—}H \\ CH_3 \end{array} \qquad \begin{array}{c} CO_2H \\ H_2N\text{—}\!\!\!+\!\!\!\text{—}H \\ CH_2.S.S.CH_2.CH(NH_2).CO_2H \end{array}$$

(10) 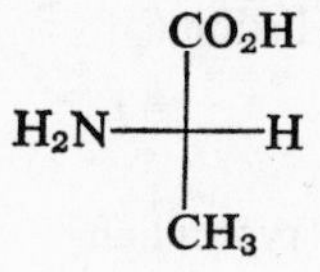 **(11)** 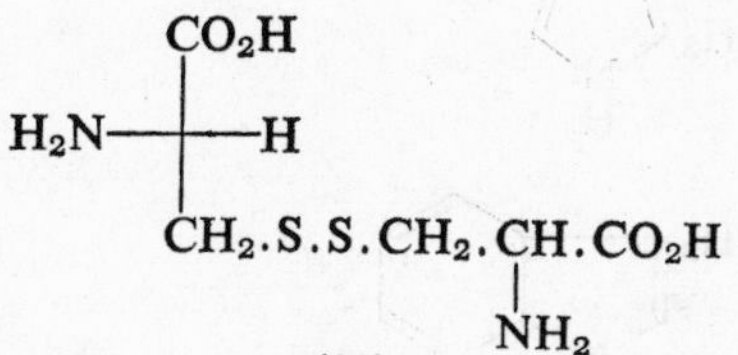 **(12)**

Nomenclature

Peptides are conventionally written with the terminal α-amino group to the left and are named as acyl substituents of the carboxyl-terminal amino acid residue. Thus, the simplest peptide, referred to above (**1**; $R^1 = R^2 =$ H), is glycylglycine. Usually, the three letter symbols given in Table 1.1

Table 1.1. Amino acids ($NH_2.CHR.CO_2H$) commonly obtained by the hydrolysis of proteins. The imino acid proline, a common constituent of proteins, has the structure HN—$CH.CO_2H$ (ring) It is abbreviated as Pro. Some proteins contain residues of other amino acids; for example, collagen, which is one of the most abundant proteins, contains residues of hydroxyproline and hydroxylysine

R	Name	Abbreviation
H	Glycine	Gly
Me	Alanine	Ala
$CH.Me_2$	Valine	Val
$CH_2.CH.Me_2$	Leucine	Leu
$CH.MeEt$	Isoleucine	Ile
$(CH_2)_4NH_2$	Lysine	Lys
$(CH_2)_3.NH.C(=NH).NH_2$	Arginine	Arg
$CH_2.CO_2H$	Aspartic acid	Asp
$CH_2.CO.NH_2$	Asparagine	Asn
$(CH_2)_2.CO_2H$	Glutamic acid	Glu
$(CH_2)_2.CO.NH_2$	Glutamine	Gln
CH_2OH	Serine	Ser
$CHOH.Me$	Threonine	Thr
$(CH_2)_2.S.Me$	Methionine	Met
$CH_2.SH$	Cysteine	Cys
$CH_2.S.S.CH_2.CH(NH_2).CO_2H$	Cystine	Cys | Cys
$CH_2.C_6H_5$	Phenylalanine	Phe
$CH_2.p\text{-}C_6H_4OH$	Tyrosine	Tyr
CH_2 (imidazole ring: N, NH)	Histidine	His
CH_2 (indole ring: NH)	Tryptophan	Trp

also refer to structures written with the free α-amino group to the left. For example, 'Gly' is $NH_2CH_2CO_2H$ and not $HO_2CCH_2NH_2$. A dot or hyphen before the symbol indicates substitution for a hydrogen of the amino group ($.NHCH_2CO_2H$); after the symbol, substitution in place of the hydroxyl of the carboxyl group ($NH_2CH_2CO.$); and above or below the symbol, substitution, as appropriate, in the side chain (for example

|
Glu means $NH_2CH(CH_2CH_2CO.)CO_2H$. Glycylglycine (**1**; $R^1 = R^2 =$ H) is therefore abbreviated to 'Gly.Gly'; arginylvalyltyrosine (**13**), for example, to 'Arg.Val.Tyr'; β-aspartylleucine (**14**) to Asp (with ┌Leu above). If the sequence of the residues in a peptide is not known, the symbols of the constituent amino acids are separated by commas. 'Gly,Ala' could therefore be alanylglycine (**15**) or glycylalanine (**16**). In the abbreviated structures of cyclic peptides, it is often necessary to write residues in the reverse form. In this case, arrows are sometimes used to indicate the direction of the amide bonds (CO → NH). Thus, cyclodi(valylornithylleucylphenylalanylprolyl) (**17**) is abbreviated to:

Val.Orn.Leu.Phe.Pro
Pro.Phe.Leu.Orn.Val   or   (Val.Orn.Leu.Phe.Pro)$_2$

HN=C(NH$_2$)–NH–(CH$_2$)$_3$–; Me Me CH; OH-C$_6$H$_4$-CH$_2$

$NH_2.CH.CO.NH.CH.CO.NH.CH.CO_2H$

(**13**)

CHMe$_2$
|
CH$_2$
|
$CH_2.CO.NH.CH.CO_2H$
|
$NH_2.CH.CO_2H$

(**14**)

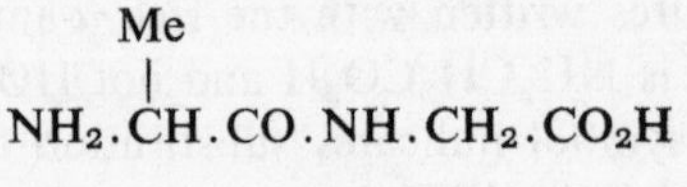

$$\begin{array}{c} Me \\ | \\ NH_2.CH.CO.NH.CH_2.CO_2H \end{array}$$

(15)

$$\begin{array}{r} Me \\ | \\ NH_2.CH_2.CO.NH.CH.CO_2H \end{array}$$

(16)

NH_2 $\quad$ $CHMe_2$

$CHMe_2$ $\quad$ $(CH_2)_3$ $\quad$ CH_2 $\quad$ CH_2 (phenyl above CH_2; pyrrolidine ring)

$NH.CH.CO.NH.CH.CO.NH.CH.CO.NH.CH.CO.N—CH.CO$

$CO.CH—N.CO.CH.NH.CO.CH.NH.CO.CH.NH.CO.CH.NH$

CH_2 $\quad$ CH_2 $\quad$ $(CH_2)_3$ $\quad$ $CHMe_2$

$CHMe_2$ $\quad$ NH_2

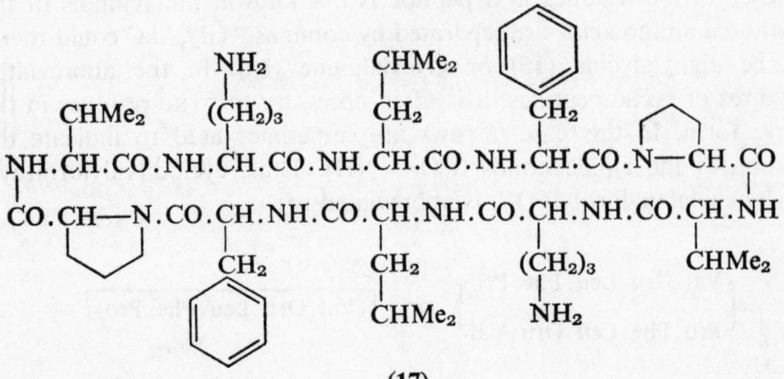

(17)

Exercise 1.7

The structure of the pituitary hormone, vasopressin, is abbreviated to:

Cys.Tyr.Phe
Cys.Asn.Gln
Pro.Arg.Gly.NH_2

or

Cys.Tyr.Phe.Gln.Asn.Cys.Pro.Arg.Gly.NH_2

Draw out its full covalent structure.

CHAPTER TWO

Determination of Primary Structure

General considerations

In the last chapter, we saw that it is difficult or even impossible, except in the simplest cases, to express the composition of a peptide meaningfully in terms of its elemental analysis. This disadvantage is more than counterbalanced by the fact that the proportions of the different amino acid residues in the peptide can be determined. An elemental analysis tells little of the arrangement of the constituent atoms in a molecule; amino acid analysis, since it deals with units which retain many of the covalent bonds of the original molecule, tells a great deal about it. The total covalent structure of the peptide is merely an expression of the sequence of the individual amino acid residues in the peptide chain. This sequence is referred to as the primary structure of the peptide.

*Exercise 2.1**

(a) A pentapeptide Gly_1, Glu_2, Val_1, Ile_1, obtained by the partial hydrolysis of insulin, gave, on further hydrolysis, five peptides of the following compositions: (i) Glu_1, Gly_1, Ile_1, Val_1; (ii) Glu_1, Ile_1, Val_1; (iii) Glu_1, Val_1; (iv) Glu_2; (v) Ile_1, Val_1. Deduce the two possible primary structures for the pentapeptide which would be in accord with this evidence and consider how they might be distinguished experimentally.

(b) Carry out the same exercise for the octapeptide, Asp_1, Glu_2, Leu_2, Ser_1, Tyr_2, which was obtained from insulin in the same investigation. On hydrolysis, it gave the following peptides: (i) Leu_1, Tyr_1; (ii) Asp_1, Glu_1; (iii) Glu_1, Tyr_1; (iv) Leu_1, Ser_1, Tyr_1; (v) Glu_2, Leu_1; (vi) Asp_1, Glu_1, Leu_1; (vii) Asp_1, Tyr_1; (viii) Glu_1, Leu_1, Ser_1, Tyr_1.

Whereas amino acid analysis necessitates the cleavage of all peptide bonds, i.e. total hydrolysis, residue sequence determination requires that some peptide bonds remain intact, i.e. partial hydrolysis. If a sufficient number of smaller peptides is obtainable by partial hydrolysis, the sequence of the amino acid residues in the parent peptide can be deduced from the amino acid analyses of these fragments. This is illustrated in Exercise 2.1*, where it can be seen that, excluding rearrangements during the process of partial hydrolysis, the sequence of the residues in the pentapeptide must be Gly.Ile.Val.Glu.Glu, or its reverse; and of the residues in the octapeptide, Ser.Leu.Tyr.Glu.Leu.Glu.Asp.Tyr, or its reverse. There is insufficient evidence to decide which way round the sequence should be read, but the identification of one of the terminal residues, the *N*-terminal or the *C*-terminal residue, would be sufficient to resolve this ambiguity.

Generally, it is not easy to isolate a sufficient number of sub-peptides for sequence studies to be based solely upon amino acid analysis in this way. Any additional information about the derived peptides is therefore valuable in that it reduces the number of peptides which has to be isolated. For example, if it were known that peptide (ii) in Exercise 2.1(a) was Ile. Val.Glu, it would not have been essential to isolate peptides (iii) and (v). Methods of identifying the terminal residues of peptides and methods which indicate, by stepwise degradation, the terminal residue sequences, have both been evolved. The importance of basing sequences on a multiplicity of information cannot be overstressed, since many peptide molecules are so large that the confirmation of a proposed structure by synthesis is exceedingly difficult.

Partial hydrolysis of peptides

The approach to sequence determination which is outlined above has three fundamental requirements; it must be possible (i) to arrest the hydrolysis of the parent peptide so that it does not proceed all the way to amino acids; (ii) to avoid the rearrangement of residue sequences during the partial hydrolysis; and (iii) to obtain peptides which represent overlapping regions of the original peptide chain. Partial hydrolysis under acidic conditions is frequently employed in sequence studies. To see how it meets these various requirements, it is necessary to consider the mechanism of the hydrolysis reaction in somewhat greater detail.

*Exercise 2.2**

Account qualitatively for the following data which relate to the hydrolysis of the indicated peptide bonds in 2M hydrochloric acid.

Peptide A B C	k^a_{A-B}	ΔH^b_{A-B}	ΔS^c_{A-B}	k_{A-BC}	ΔH_{A-BC}	ΔS_{A-BC}	k_{AB-C}	ΔH_{AB-C}	ΔS_{AB-C}
Gly.Gly —	1·46	20·3	−24·0	—	—	—	—	—	—
Gly.Gly.Gly	—	—	—	1·46	20·62	−21·34	5·51	20·37	−19·15
Leu.Gly —	0·27	21·6	−23.4	—	—	—	—	—	—
Leu.Gly.Leu	—	—	—	0·345	21·0	−23·5	3·12	20·0	−22·2
Gly.Leu —	0·61	19·8	−27·1	—	—	—	—	—	—
Gly.Leu.Gly	—	—	—	0·70	19·6	−26.0	1·30	20·2	−23·2

[a] (×10³/min) measured at 74·7°C; [b] kcal/mole; [c] cal/degree/mole.

The data quoted in Exercise 2.2* illustrate the quite general finding that, other factors being equal, the peptide bond adjacent to the terminal amino group in a peptide is the most resistant to hydrolysis under acidic conditions. In triglycine, this bond is hydrolysed at approximately the same rate as the peptide bond in glycylglycine, whereas the other peptide bond in triglycine is hydrolysed approximately 3·8 times as fast. Similarly, the *N*-terminal peptide bonds in leucylglycylleucine and glycylleucylglycine are hydrolysed approximately 1·31 and 1·15 times as fast, respectively, as the corresponding dipeptides; whereas the corresponding ratios for the *C*-terminal bonds as 5·1 and 4·8. Thermochemically, the bonds are all rather similar; ΔH values for the bond between two particular residues is remarkably constant regardless of the relative position of that bond in the peptide.

Differences in the reactivity of a given bond in various situations seem to be accountable for almost entirely in terms of an entropy effect. The slower hydrolyses involve relatively greater negative entropy changes. It is not difficult to visualize in qualitative terms why this should be so. Under acidic conditions, the terminal amino group of a peptide will be protonated.

In consequence, protonation of the neighbouring peptide bond, a necessary prerequisite to hydrolysis, and nucleophilic addition to the protonated bond will be less favoured than similar changes to peptide bonds further removed from the terminal charge.

The net effect is that partial hydrolysis of a peptide does tend to produce smaller peptides rather than amino acids. In practice, the hydrolysis is usually brought about by exposing the peptide to constant boiling or concentrated hydrochloric acid for short periods, sometimes at elevated temperatures. Very few examples of residue sequence rearrangements under these conditions have been observed, but in weaker acid concentrations, rearrangements do take place. Thus, leucyltyrosine (**18**) when treated with 2M hydrochloric acid at 80–110°C undergoes inversion of the residue sequence to tyrosylleucine (**19**) and this occurs approximately 3 to 4 times as rapidly as hydrolysis. Inversion proceeds by way of the corresponding 2,5-dioxopiperazine (**20**), and the formation of this derivative is the rate determining step. It is calculated in this case that ΔH for the piperazine-forming reaction is about 5 kcal per mole less than ΔH for the hydrolysis.

$NH_2.CH(CH_2CHMe_2).CO.NH.CH(CH_2C_6H_4OH).CO_2H$ (**18**)

$NH_2.CH(CH_2C_6H_4OH).CO.NH.CH(CH_2CHMe_2).CO_2H$ (**19**)

(**20**) 2,5-dioxopiperazine: HN, CO, OC, NH ring bearing $CH_2C_6H_4OH$ and CH_2CHMe_2 substituents

Other rearrangements sometimes occur during the hydrolysis process, but usually, these do not result in sequence alterations. Examples of such rearrangements are provided by aspartyl and asparaginyl peptides (**21**; R = OH, NH_2, OCH_3 etc.) which rearrange through a succinimide intermediate (**22**) to mixtures of α and β-aspartyl (**23**) peptides. Glutamyl and glutaminyl peptides (**24**; R = OH, NH_2, OCH_3 etc.) will rearrange in a similar manner via a glutarimide intermediate (**25**), but can also form pyrrolidone carboxylic acid derivatives (**26**). In these transformations, the nitrogen of the peptide bond, or, in the latter case, the α-amino group, serves as the nucleophile in a reaction which otherwise resembles the hydrolysis process. These rearrangements occur most readily under alkaline conditions but have also been observed under acidic conditions.

$$NH_2.CH(CH_2.CO.R).CO.NH.CHR^1.CO.R^2 \quad \textbf{(21)}$$

$$\rightleftharpoons$$

$$NH_2.CH.CO(CH_2.CO)\!>\!N.CHR^1.CO.R^2 \quad \textbf{(22)} \rightleftharpoons NH_2.CH(CH_2.CO.NH.CHR^1.CO.R^2).CO_2H + NH_2.CH(CH_2.CO_2H).CO.NH.CHR^1.CO.R^2 \quad \textbf{(23)}$$

$$NH_2.CH(CH_2.CH_2.CO.R).CO.NH.CHR^1.CO.R^2 \quad \textbf{(24)} \rightleftharpoons HN\text{—}C(=O)\text{—}CH_2\text{—}CH_2\text{—}CH.CO.NH.CHR^1.CO.R^2 \quad \textbf{(26)}$$

$$\rightleftharpoons$$

$$NH_2.CH\text{—}CH_2\text{—}CH_2\text{—}CO\text{—}N(CHR^1.CO.R^2)\text{—}CO \quad \textbf{(25)} \rightleftharpoons NH_2.CH(CH_2.CH_2.CO.NH.CHR^1.CO.R^2).CO_2H + NH_2.CH(CH_2.CH_2.CO_2H).CO.NH.CHR^1.CO.R^2$$

---

*Exercise 2.3**

Account for the following observations which relate to acidic conditions:

(a) Glycylvaline is hydrolysed much more rapidly than valylglycine.

(b) Glycylalanine is hydrolysed at approximately the same rate as alanylglycine, but much more slowly than glycylserine.

Allowing for differences related to molecular weight, it might be predicted on the basis of the mechanism so far considered, that peptide bonds between the various types of amino acid residues should be hydrolysed at much the same rate. Two factors, steric hindrance and the possible participation of neighbouring groups in the hydrolytic step, have not been taken into account; either can be important in specific cases. In Exercise 2.3(a)*, for example, the resistance to hydrolysis of a peptide bond involving the carboxyl group of valine is illustrated. Presumably, this is a steric effect, in which the approach of the water molecule to the protonated amide bond is hindered by the bulky isopropyl side-chain of the valine residue. In Exercise 2.3(b)*, the effect of neighbouring group participation is encountered because, in *N*-acyl serine derivatives (**27**), the β-hydroxyl group of serine is ideally situated for nucleophilic attack on the protonated amide. The hydroxyoxazolidone (**28**), so formed, undergoes ring opening to the corresponding β-ester (**29**) and, since simple esters usually hydrolyse more rapidly than amides, the cleavage is complete. The transformation (**27** → **29**) is referred to as an N → O acyl shift. Under mildly alkaline conditions, the ester rearranges to the acyl serine derivative (O → N acyl shift).

$$\mathrm{R.CHR^1.\underset{\underset{O}{\|}}{C}.NH.CH(CH_2OH).CO.R^2} \quad \textbf{(27)}$$

$$\Big\Updownarrow \mathrm{H^+}$$

$$\left.\begin{array}{l} \mathrm{R.CHR^1.\underset{\underset{O}{\|}}{C}.\overset{+}{N}H_2.CH(CH_2OH).CO.R^2} \\ \Big\Updownarrow \\ \mathrm{R.CHR^1.\underset{\underset{^+OH}{\|}}{C}.NH.CH(CH_2OH).CO.R^2} \end{array}\right\} \rightleftharpoons \mathrm{R.CHR^1.C(OH)} \text{—} \overset{+}{\mathrm{N}}\mathrm{H_2}\text{—}\mathrm{CH(CO.R^2)}\text{—}\mathrm{CH_2}\text{—}\mathrm{O}\text{— (ring)} \quad \textbf{(28)}$$

$$\Big\Updownarrow$$

$$\mathrm{R.CHR^1.CO_2H} + \mathrm{\overset{+}{N}H_3.CH(CH_2OH).CO.R^2} \rightleftharpoons \mathrm{R.CHR^1.\underset{\underset{O}{\|}}{C}.O.CH_2.CH(\overset{+}{N}H_3).CO.R^2} \quad \textbf{(29)}$$

Most peptide bonds are hydrolysed at comparable rates under acidic conditions so that cleavage is usually random and overlapping sequences are produced. However, because some bonds are recalcitrant and others

Reaction with 2,4-dinitrofluorobenzene is carried out with a weakly alkaline aqueous solution of the peptide. Small amounts of alcohol may be added to increase the solubility of the 2,4-dinitrofluorobenzene, but the reaction mixture is often heterogeneous. Excess reagent is removed from the alkaline mixture at the end of the reaction by ether extraction. Because the basicity of the amino group is suppressed, the DNP-peptide is no longer a zwitterion and, after acidification, it can often be extracted into an organic solvent. Similarly, after hydrolysis of the DNP-peptide, the DNP-derivative of the *N*-terminal amino acid is usually separated easily from the amino acid salts, for example, by extracting it into ethyl acetate. In some instances, due to the presence of side chain functional groups, DNP-peptides and DNP-amino acids cannot be extracted in this way and a different procedure is followed. In general, DNP-amino acids are identified by chromatography and are easily detected and estimated by their yellow colour ($\lambda_{max} \approx 360$ nm, $\epsilon \approx 16{,}000$).

---

*Exercise 2.5**

Write a detailed mechanism for the initial reaction of the peptide with 2,4-dinitrofluorobenzene. Would you expect reaction with 3,5-dinitrofluorobenzene to take place as readily? How about reaction with 2,4-dinitrochlorobenzene? 2,4-Dinitrophenol is usually a contaminant of the DNP-peptide. How does it arise; how could it be detected; and how might it be separated from the DNP-peptide?

The initial reaction of the peptide with 2,4-dinitrofluorobenzene is a typical aromatic nucleophilic substitution reaction. Bear in mind that aromatic substitution in the benzene ring is usually an electrophilic process; it is the strong electron-withdrawing effect of the nitro groups ($-I, -M$) which facilitates the nucleophilic process. It does this, of course, by extending charge delocalization and so stabilizing the negatively-charged intermediate (**35**; R = $CHR^1CONHCHR^2CO$... etc.). If the substituents were in the 1,3,5 positions, only ring delocalization would be possible and the influence of the nitro groups would be limited to the inductive effect.

i.e.

(35)

2,4-Dinitrochlorobenzene can be used to label amino groups but in this case, more vigorous conditions are necessary to drive the reaction to completion and the danger of side-reactions is therefore greater. It will be recalled that the reactivity of derivatives of the halogens in nucleophilic displacement reactions differs characteristically in the aliphatic and aromatic series. The rates of S_N reactions usually increase in the series $F \ll Cl < Br < I$, whereas for $S_N$ar reactions, the series is $F \gg$ Cl, Br, I. In the aliphatic series, the *C*-halogen bond-strengths are important and the rate determining step is the cleavage of this bond; in the aromatic series, formation of the charged intermediate is usually rate-determining.

2,4-Dinitrophenol is usually present in the DNP-amino acid extract. It is presumably formed by competitive hydrolysis during the reaction of the peptide with 2,4-dinitrofluorobenzene. The dinitrophenol can be distinguished from DNP-amino acids on chromatograms because it is decolourised on exposure to acid vapour (p-$NO_2C_6H_4O^-$, $\lambda_{max} = 400$ nm, $\epsilon = 15,000$; p-$NO_2C_6H_4OH$, $\lambda_{max} = 320$ nm, $\epsilon = 9000$). However, when it is present in considerable amounts, dinitrophenol can interfere with the chromatographic separation and in this case it can be removed by sublimation under reduced pressure. It might be thought that the reactivity of the dinitrochlorobenzene derivative could be enhanced by further substitution to the stage at which it would undergo aminolysis by the peptide at room temperature. Indeed, 2,4,6-trinitrochlorobenzene is more reactive than the dinitro compound and has been used for terminal residue identification, but it suffers from the disadvantage that it is very readily hydrolysed to picric acid.

2,4-Dinitrophenyl amino acids are slowly degraded on exposure to light to give varying proportions of 4-nitro-2-nitrosoaniline (**36**) and 2-substituted-6-nitrobenzimidazole-1-oxides (**37**), dependent on the prevailing pH.

NH.CHR.CO_2H, NO_2, NO_2 — $h\nu$, $-CO_2$, $-RCHO$ → NH_2, NO, NO_2 (**36**)

— $h\nu$, $-CO_2$, $-H_2O$ → NH, C—R, O_2N, $\overset{+}{N}$, O^- (**37**)

Degradation is not pronounced under the usual conditions of end-group determination, but it is necessary to be cautious on this account, especially during chromatographic processes when the material is spread over a large surface.

*Exercise 2.6**

p-^{131}Iodobenzene sulphonyl chloride reacts with compounds which possess an amino group to yield *N*-*p*-^{131}iodobenzene sulphonyl (pipsyl) derivatives. The S–N bond in these derivatives is resistant to hydrolysis and this type of labelling has been used to identify *N*-terminal amino acid residues. Account for the stability shown by *N*-pipsyl derivatives under acidic conditions.

A method which is even more sensitive than the 2,4-dinitrofluorobenzene technique for *N*-terminal residue identification involves reacting the peptide with 1-dimethylaminonaphthalene-5-sulphonyl chloride (dansyl chloride) (**38**) under mildly basic conditions to form the corresponding dansyl peptide (**39**). Hydrolysis of the dansyl derivative yields the *N*-dansyl derivative of the terminal residue (**40**) which can be identified chromatographically and which is easily detected by its fluorescence in ultraviolet light. This method of detection is about one hundred times more sensitive than the detection of DNP-amino acids.

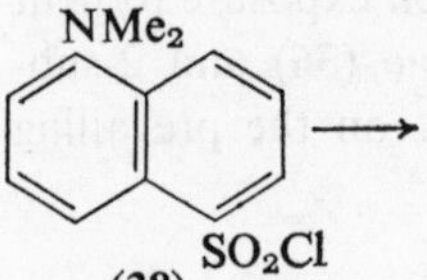

NMe_2

SO_2Cl

(**38**)

NMe_2

$SO_2.NH.CHR^1.CO.NH.CHR^2.CO\ldots NH.CHR^n.CO_2^-$

(**39**)

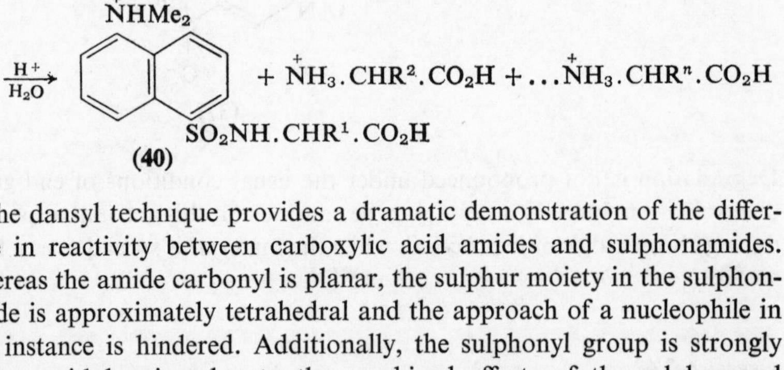

$\overset{+}{N}HMe_2$

$\xrightarrow[H_2O]{H^+}$ $+ \overset{+}{N}H_3.CHR^2.CO_2H + \ldots \overset{+}{N}H_3.CHR^n.CO_2H$

$SO_2NH.CHR^1.CO_2H$

(**40**)

The dansyl technique provides a dramatic demonstration of the difference in reactivity between carboxylic acid amides and sulphonamides. Whereas the amide carbonyl is planar, the sulphur moiety in the sulphonamide is approximately tetrahedral and the approach of a nucleophile in this instance is hindered. Additionally, the sulphonyl group is strongly electron-withdrawing due to the combined effects of the sulphur and oxygen atoms and, because of this effect, sulphonamides are appreciably acidic. In alkaline solution, they lose a proton from the nitrogen atom, a characteristic which is exploited, of course, in the Hinsberg method for separating amines. Not unrelated is the fact that sulphonic acids are strong acids, comparable in strength to sulphuric acid, whilst even a CH_2 or CH group can be strongly acidic when substituted with sulphonyl groups. It follows that protonation of sulphonamides will not readily occur and, even if this necessary preliminary to nucleophilic attack is achieved, approach of the nucleophilic water molecule to the sulphur will be subject to steric hindrance. Consequently, only very vigorous acidic conditions, for example, concentrated hydrochloric acid at 200°C, lead to fission of the S–N bond. Hydrolysis under alkaline conditions is even more difficult, presumably because the SO_2–N system has high electron density. These considerations account for the stability of dansylamino acids, and of the older and lesser used pipsyl derivatives (Exercise 2.6*), under conditions in which peptide bonds are hydrolysed. The SO_2–N bond can be cleaved by sodium–liquid ammonia reduction without affecting peptide bonds; this type of cleavage is discussed further in Chapter 3.

---

*Exercise 2.7**

Peptide bonds are cleaved by hot anhydrous hydrazine. What is the nature of this reaction? Is it likely to form the basis of a technique for terminal residue identification?

There are no methods for the identification of *C*-terminal residues which are as effective and hence as important as the methods discussed above for the identification of *N*-terminal residues, but one particular *C*-terminal method is noteworthy. The hydrazinolysis (**41** → **42**) of peptides (Exercise 2.7*) depends on nucleophilic attack by the hydrazine molecule on the amide bond. Since hydrazine is basic as well as nucleophilic the *C*-terminal residue of the peptide will form a salt. Work-up of the reaction mixture will therefore yield the free *C*-terminal amino acid, whilst all other residues will be present as hydrazides. In this way, the *C*-terminal residue of a peptide can be identified.

$$NH_2.CHR^1.CO.NH.CHR^2.CO...NH.CHR^n.CO_2^- \xrightarrow{NH_2NH_2}$$

(41)

$$NH_2.CHR^1.CO.NH.NH_2 + NH_2.CHR^2.CO.NH.NH_2... + NH_2.CHR^n.CO_2^-$$

(42)

Exercise 2.8

It is interesting to speculate how the different products of hydrazinolysis might be separated. Could one distinguish asparagine, aspartic acid, glutamine and glutamic acid residues in the original peptide by identifying the products of hydrazinolysis?

First thoughts might indicate that the identification of a terminal residue can only confirm the findings of stepwise degradation so that, apart from the check which it provides, terminal residue identification is largely superfluous. This is not the case. Terminal residue identification can be very informative and often sequence methods can only be applied with confidence if the terminal residue has first been identified. For example, during the 2,4-dinitrophenylation technique, if non-stoichiometric amounts of DNP-amino acids were formed, it might mean that the peptide was impure; if equimolar amounts of more than one DNP-amino acid were formed, the preparation could be an equimolar mixture of different peptides, or, more likely, it could consist of a peptide which possessed more than one chain, joined perhaps by the disulphide bonds of cystine. In the former case, further purification would be necessary before sequence studies could be started; in the latter, the peptide chains would first need to be separated. Three methods which are used to cleave disulphide-containing compounds (**43**) involve, respectively, oxidation to sulphonic (cysteic) acid derivatives (**44**), oxidative sulphitolysis to the *S*-sulphonate derivatives (**45**), and reductive alkylation to thioether derivatives (**46**).

$$R^1.NH.CH(CH_2.S.S.CH_2).CO.R^2 \ / \ R^3.NH.CH.CO.R^4 \quad (\mathbf{43})$$

$$\xrightarrow{HCO_3H} R^1.NH.CH(CH_2SO_3H).CO.R^2 + R^3.NH.CH(CH_2SO_3H).CO.R^4 \quad (\mathbf{44})$$

$$\xrightarrow{SO_3^{2-}} R^1.NH.CH(CH_2.S.SO_3^-).CO.R^2 + R^3.NH.CH(CH_2.S^-).CO.R^4 \quad (\mathbf{45})$$

$$+ R^1.NH.CH(CH_2.S^-).CO.R^2 + R^3.NH.CH(CH_2.S.SO_3^-).CO.R^4 \quad (\mathbf{45})$$

$$\xrightarrow{(i)\ Na/NH_3;\ (ii)\ Ph.CH_2.Cl} R^1.NH.CH(CH_2.S.CH_2.Ph).CO.R^2 + R^3.NH.CH(CH_2.S.CH_2.Ph).CO.R^4 \quad (\mathbf{46})$$

Terminal sequence determination

It will be apparent by now that the peptide bond is by no means inert and that, in appropriate circumstances, both the carbonyl and imino groups are available for reaction. Once this has been realized, attempts to devise a technique which will permit the removal of either the *N* or *C*-terminal residues in a stepwise manner can be made more hopefully.

---

*Exercise 2.9**

By considering the N → O acyl shift of Exercise 2.3(b)*, outline a series of reactions which might form the basis of a technique for *C*-terminal determination.

N → O Acyl migration in serine-containing peptides is an example of a reaction which involves acid catalysed nucleophilic addition to the amide carbonyl group. Since, in principle, the terminal carboxyl group of a peptide can be reduced to the primary alcohol (**47**), N → O acyl migration could form the basis of a technique for *C*-terminal sequence identification.

The *C*-terminal residues would be identified in a stepwise manner as the corresponding β-amino alcohols (**48**).

$$\mathrm{R.CO.NH.CHR^1.CO\ldots NH.CHR^{n-1}.CO.NH.CHR^n.CO_2H}$$

$$\downarrow \text{MeOH/HCl}$$

$$\mathrm{R.CO.NH.CHR^1.CO\ldots NH.CHR^{n-1}.CO.NH.CHR^n.CO_2Me}$$

$$\downarrow \text{LiH}$$

$$\mathrm{R.CO.NH.CHR^1.CO\ldots NH.CHR^{n-1}.CO.NH.CHR^n.CH_2OH} \quad \textbf{(47)}$$

$$\downarrow \text{SOCl}_2$$

$$\mathrm{R.CO.NH.CHR^1.CO\ldots NH.CHR^{n-1}.CO_2.CH_2.CHR^n.\overset{+}{N}H_3Cl^-}$$

$$\downarrow \text{LiH}$$

$$\mathrm{R.CO.NH.CHR^1.CO\ldots NH.CHR^{n-1}.CH_2OH + HO.CH_2.CHR^n.NH_2} \quad \textbf{(48)}$$

This reaction sequence is quoted merely to indicate the possibilities. In fact, it does not provide a satisfactory method because the reactions do not go to completion and side-reactions occur. It is important in sequence work that reactions should proceed quantitatively. This is a difficult requirement to meet and, of the many possible chemical approaches which have been investigated, only one, the Edman technique for *N*-terminal sequence determination, is in common use.

In the Edman degradation, the peptide is first condensed with phenylisothiocyanate under mildly alkaline conditions (pH $\approx$ 8) to give the phenylthiocarbamyl derivative (**49**).

$$\text{Ph—N=C=S} \quad + \quad \text{H}_2\text{N.CHR}^1\text{.CO.NH.CHR}^2\text{.CO...NH.CHR}^n\text{.CO}_2^-$$

$$\downarrow$$

$$\text{Ph.NH.}\underset{\underset{\text{S}}{\|}}{\text{C}}\text{.NH.CHR}^1\text{.CO.NH.CHR}^2\text{.CO...NH.CHR}^n\text{.CO}_2^-$$

λ_{max} = 245 nm (EtOH)

(49)

Under acidic conditions, the phenylthiocarbamyl derivative undergoes what amounts to an N $\rightarrow$ S acyl shift (**50** $\rightarrow$ **51**).

$$\textbf{(49)} \underset{}{\overset{\text{H}^+}{\rightleftharpoons}} \text{Ph.NH.C}(=\text{S})\text{—}\ddot{\text{N}}\text{H—CHR}^1\text{—C}=\overset{+}{\text{O}}\text{H} \text{ (—NH.CHR}^2\text{.CO...NH.CHR}^n\text{.CO}_2\text{H)} \longrightarrow$$

(50)

$$\text{Ph.NH.C}\langle\overset{+}{\text{N}}\text{H—CHR}^1\text{—C(=O)—S}\rangle + \overset{+}{\text{N}}\text{H}_3\text{.CHR}^2\text{.CO...NH.CHR}^n\text{.CO}_2\text{H}$$

(51) λ_{max} = 230 nm (EtOH)

There are several analogies for this reaction (e.g. **52** $\rightarrow$ **53**; **54** $\rightarrow$ **55**).

$$\text{Ph.NH.}\overset{\overset{\text{S}}{\|}}{\text{C}}\text{.NH}_2 \xrightarrow{\text{AcCl}} \text{Ph.NH—}\overset{\overset{\text{S.CO.CH}_3}{|}}{\text{C}}\text{=}\overset{+}{\text{N}}\text{H}_2\ \text{Cl}^- \rightleftharpoons \text{Ph.}\overset{\overset{\text{CO.CH}_3}{|}}{\text{N}}\text{—CS—NH}_2 + \text{HCl}$$

(52) **(53)**

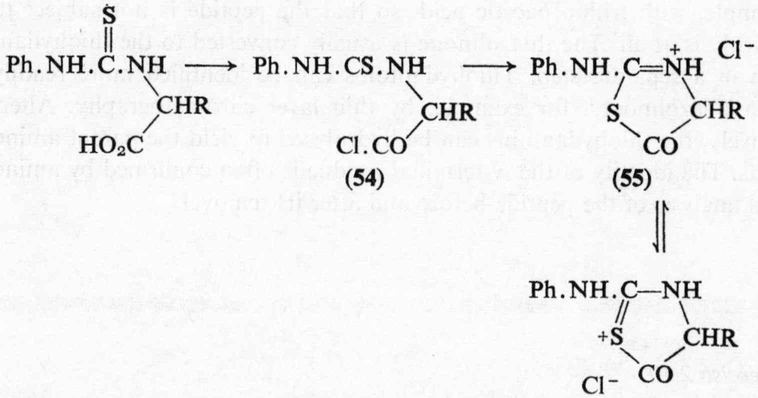

$$\text{Ph.NH.C(=S).NH.CHR.CO}_2\text{H} \longrightarrow \text{Ph.NH.CS.NH.CHR.CO.Cl}\ \textbf{(54)} \longrightarrow \text{Ph.NH.C}{=}\overset{+}{\text{N}}\text{H}\ \text{Cl}^-\ \textbf{(55)} \rightleftharpoons \text{Ph.NH.C—NH},\ {+}\text{S},\ \text{Cl}^-$$

However, the first product of the Edman cleavage, the 4-substituted-2-anilinothiazolinone (**51**) is unstable and it is converted in aqueous solution, via the phenylthiocarbamylamino acid (**56**), to the isomeric 5-substituted-3-phenyl-2-thiohydantoin (**57**). The intermediate thiazolinone was therefore overlooked for a time and the nature of the reaction was not entirely understood. Under anhydrous conditions, the intermediate thiazolinone derivative probably undergoes a slow direct rearrangement to the corresponding hydantoin (**51** → **57**).

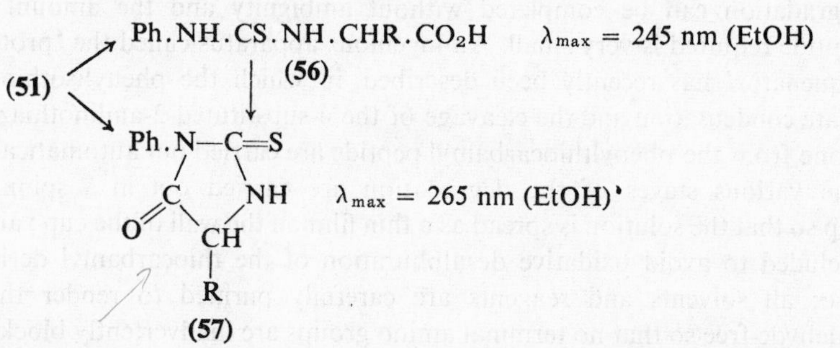

$$\textbf{(51)} \rightarrow \text{Ph.NH.CS.NH.CHR.CO}_2\text{H}\ \textbf{(56)} \quad \lambda_{max} = 245\ \text{nm (EtOH)}$$

$$\textbf{(51)} \rightarrow \text{Ph.N—C=S, C(=O), NH, CH, R}\ \textbf{(57)} \quad \lambda_{max} = 265\ \text{nm (EtOH)}$$

The realization that the reaction occurs in several distinct steps has far-reaching practical significance. Originally, the rate of formation of the thiohydantoin derivative was taken as a measure of the cleavage reaction. Now, it is appreciated that, at acid concentrations greater than approximately 0·2M, the initial cleavage reaction (**49** → **51**) takes place more rapidly than hydantoin formation (**51** → **56** → **57**), whilst the hydrolysis of the thiazolinone, (**51** → **56**) is actually inhibited in strongly acidic solution. It is therefore best to isolate the thiazolinone and thereby avoid exposing the peptide unnecessarily to hydrolysing conditions. Formation of the thiazolinone can be carried out under anhydrous conditions, for example, with trifluoroacetic acid, so that the peptide is not subject to hydrolysis at all. The thiazolinone is usually converted to the thiohydantoin in a separate step. Thiohydantoins can be identified more readily than thiazolinones, for example, by thin layer chromatography. Alternatively, the thiohydantoins can be hydrolysed to yield the parent amino acids. The identity of the *N*-terminal residue is often confirmed by amino acid analysis of the peptide before and after its removal.

Exercise 2.10

Phenylthiocarbamylleucine reacts with phosphorus oxychloride to give a neutral compound $C_{13}H_{17}N_2OSCl$ which rearranges spontaneously to an isomeric base hydrochloride. Outline the reactions involved.

When optimal conditions are employed, many cycles of the Edman degradation can be completed without ambiguity and the amount of peptide required is very small. An ingenious apparatus called the 'protein sequenator' has recently been described, in which the phenylisothiocyanate condensation and the cleavage of the 4-substituted-2-anilinothiazolinone from the phenylthiocarbamyl peptide are carried out automatically. The various stages of the degradation are carried out in a spinning cup so that the solution is spread as a thin film on the wall of the cup; air is excluded to avoid oxidative desulphuration of the thiocarbamyl derivative; all solvents and reagents are carefully purified to render them aldehyde-free so that no terminal amino groups are inadvertently blocked. The formation of the thiazolinones is carried out under anhydrous conditions and the extracted thiazolinone derivatives are collected in a fraction collector and identified separately by converting them to the corresponding substituted phenylthiohydantoins.

With the automated machine, more than fifteen residues can be removed from a peptide every twenty-four hours and the yield per cycle is better than 98 per cent. Approximately 0·25 μmoles of the peptide are required for the degradation. The machine has been tested on a protein and the first sixty amino acid residues were correctly identified, which approximates to the theoretical limit if a 2 per cent loss is assumed at each stage of the cycle. Of course, a small increase in the efficiency at each step would extend quite considerably the length of sequence which could be determined and so even more impressive developments of this technique can be expected.

Although the formation of phenylthiohydantoins from phenylthiocarbamyl derivatives has been investigated most extensively, other derivatives of the same form (**58**; R = Ph, SH, OR) will rearrange in the same way and could form the basis of a stepwise technique.

```
     NH—CHR¹
     |     |
     C     C=O
    / \\    \
   R    S    NH.CHR²CO...NH.CHRⁿ.CO₂H
```

$$\mathrm{R{-}C({=}S){-}NH{-}CHR^1{-}C({=}O){-}NH.CHR^2CO\ldots NH.CHR^n.CO_2H}$$

(58)

Exercise 2.11

N-terminal sequence degradations have been investigated in which the terminal residue is removed from the peptide as a lactam. The peptide is first reacted with a nitrobenzene derivative, for example, 2-nitro-4-carboxymethylfluorobenzene or 2,4-dinitrofluorobenzene itself, and the nitro-group is subsequently reduced to form the nucleophile. Outline the reaction sequence and suggest appropriate conditions for the various stages of the degradations.

Only one chemical method is of value for the determination of *C*-terminal sequences and this involves the stepwise cleavage of the terminal residues as thiohydantoins (**59** → **61**).

In this technique, the intermediate acylisothiocyanate derivatives (**60**) can be obtained by reacting the acyl peptide with ammonium thiocyanate in acetic anhydride, although more sophisticated reagents which are more satisfactory have been described. Cyclization, in which the imino group of the peptide is the nucleophile, resembles the first stage of the Edman degradation. The 5-substituted-2-thiohydantoins (**61**) can usually

$$\mathrm{R.CO.NH.CHR^1.CO.NH.CHR^2.CO...NH.CHR^{n-1}.CO.NH.CHR^n.CO_2H} \quad \textbf{(59)}$$

$$\downarrow$$

$$\mathrm{R.CO.NH.CHR^1.CO.NH.CHR^2.CO...NH.CHR^{n-1}.CO.NH.CHR^n.CONCS} \quad \textbf{(60)}$$

$$\downarrow \text{warm}$$

$$\mathrm{R.CO.NH.CHR^1.CO.NH.CHR^2.CO...NH.CHR^{n-1}.CO.\overline{N{-}CHR^n.CO.NH.CS}}$$

(ring: N—CHR^n, SC, CO, NH)

$$\downarrow \text{mild alkaline hydrolysis}$$

$$\mathrm{R.CO.NH.CHR^1.CO.NH.CHR^2.CO...NH.CHR^{n-1}.CO_2^- + \overline{NH{-}CHR^n.CO.NH.CS}} \quad \textbf{(61)}$$

(ring: NH—CHR^n, SC, CO, NH)

be formed under conditions which do not cause too much damage to the residual peptide, but side reactions make the technique less satisfactory generally than the Edman degradation.

Endopeptidases, especially carboxypeptidase and various types of aminopeptidase, are sometimes used in sequence determinations. Carboxypeptidase, isolated from pancreatic tissue, cleaves the *C*-terminal peptide bond. This means that peptides are degraded by the enzyme in a stepwise manner from the *C*-terminus. Ideally, a kinetic study of the degradation would indicate the sequence of the residues, but the picture is complicated because all peptide bonds are not cleaved at the same rate. The type of difficulty this creates is easily demonstrated. Consider, for example, a peptide in which the penultimate peptide bond is cleaved much more rapidly than the terminal peptide bond. In this instance, the penultimate residue could be released at almost the same rate as the terminal residue and it would probably be impossible to distinguish which was released first. Normally, this complication severely limits the amount of sequence formation which can be obtained. The enzyme does not cleave peptide bonds involving the imino nitrogen of proline and it will only act when the terminal carboxyl group is free and the constituent amino acids possess the L-configuration. The earliest and perhaps best-known of the aminopeptidases, leucine aminopeptidase, which is isolated from kidney tissue, degrades peptides in a stepwise manner from the *N*-terminal end. Likewise, it requires the terminal (amino) group to be free and the residues to possess the L-configuration. In this instance, digestion is so rapid that it rarely provides sequence information, but digestion with leucine aminopeptidase does provide a sensitive test for the presence of D-amino acid residues.

Selective chemical cleavage

The common denominator of chemical techniques for stepwise degradation is that, in each case, the terminal residue is modified to produce a reactive neighbouring group which facilitates cleavage of the terminal peptide bond. Selective cleavage of non-terminal peptide bonds is possible when, in a similar manner, the side chain of a particular amino acid residue can be induced to participate in a cleavage reaction. This possibility provides the basis for an important alternative to partial hydrolysis for the initial fragmentation of the parent peptide.

When the molecular weight of the peptide is very large, partial acid hydrolysis often gives an impossibly complex mixture of small peptides and, even if some of the peptides can be isolated, it is extremely difficult to see how they are interrelated. Digestion with endopeptidases does help in this situation, but an unfortunate distribution of the bonds susceptible to enzymic digestion can limit the value of this approach. Nor is the specificity of enzymes rigid and it is often confusing to find peptides in these digests which possess *C*-terminal residues not normally produced by a particular enzyme, and bonds intact, which look as though they should be susceptible to cleavage.

Exercise 2.12

Outline a possible reaction sequence based on the $\alpha \rightarrow \beta$ aspartyl rearrangement (p. 14) and *C*-terminal sequence methods, which might achieve selective cleavage of a peptide at aspartic acid residues.

Exercise 2.13

Devise a scheme for the selective cleavage of peptides at serine residues.

Selective chemical cleavage at methionine and tryptophan residues, respectively, can be brought about very efficiently and the fact that these residues are relatively uncommon makes the approach particularly useful, since only limited fragmentation of the peptide results.

To bring about the methionine cleavage, the peptide is treated with cyanogen bromide. The sulphonium salt (**62**) forms and the sulphur moiety is displaced spontaneously to give the iminolactone (**63**). In aqueous acid, iminolactones hydrolyse readily and the products in this case are the amino compound and a derivative of homoserine lactone (**64**). The tryptophan cleavage also occurs via an iminolactone (**65**), formed by treating the peptide with *N*-bromosuccinimide or *N*-bromoacetamide at pH 4. Interestingly, indole-3-propano-*p*-nitroanilide is not cleaved in this way.

Me
S
CH_2
CH_2 O
$R.CO.NH.CH—C—NH.CHR^1.CO.R^2$ $\xrightarrow{CNBr}$

Me CN
S^+ + Br^-
CH_2
CH_2 O
$R.CO.NH.CH—C—NH.CHR^1.CO.R^2$ ⟶

(62)

CH_2
CH_2 O
$R.CO.NH.CH—C{=}\overset{+}{N}H.CHR^1.CO.R^2$ + MeSCNBr ⟶

(63)

CH_2
CH_2 O
$R.CO.NH.CH—C{=}O + \overset{+}{N}H_3.CHR^1.CO.R^2$

(64)

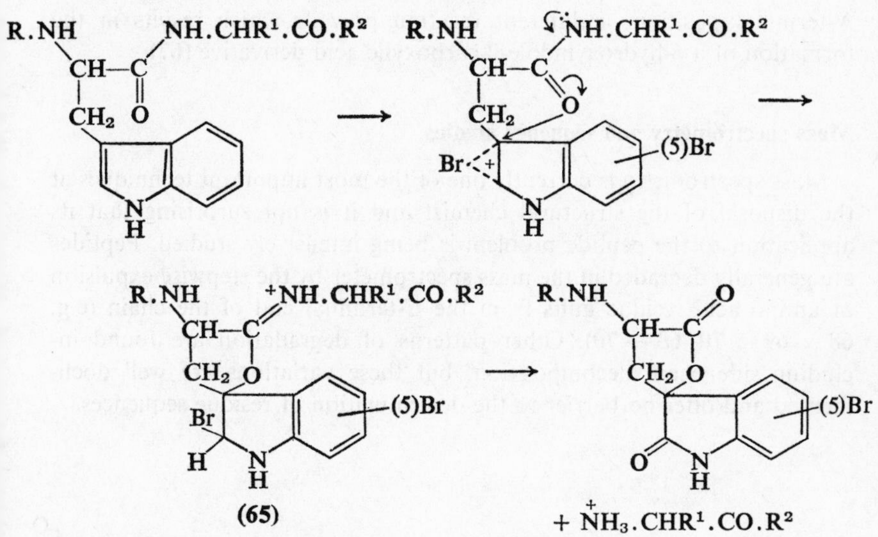

(65)

$+ \overset{+}{N}H_3.CHR^1.CO.R^2$

Exercise 2.14

Indole-3-propanoic acid, when treated with 3 moles of *N*-bromosuccinimide in methanol–acetate buffer (pH 4), gave a neutral compound as colourless needles, m.p. = 199–200°, $C_{11}H_8NO_3Br$, λ_{max}(EtOH) = 260, 308 nm, ν_{KBr} = 1779, 1736 cm^{-1}. Postulate a structure for this compound.

Peptides which contain tyrosine residues can also be cleaved by oxidative bromination. The cleavage proceeds via an intermediate iminolactone **(66)**, very like the cleavage of tryptophan peptides. When tyrosine is in the

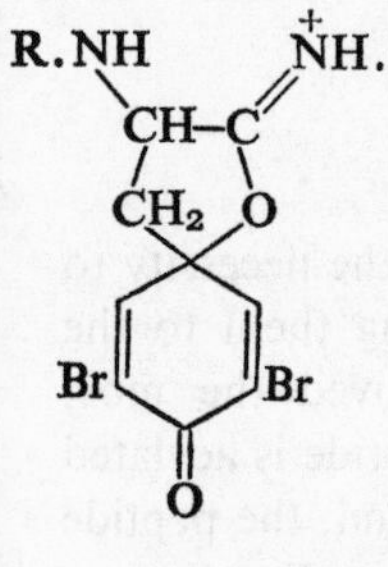

(66) 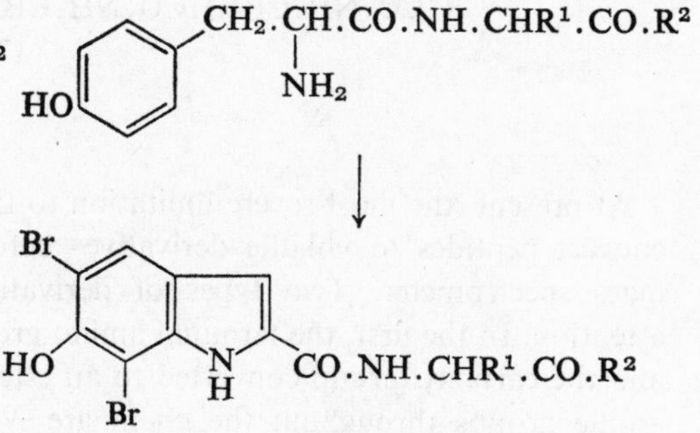 (67)

N-terminal position, a different reaction prevails which results in the formation of a 6-hydroxyindole-2-carboxylic acid derivative (**67**).

Mass spectrometry and sequence studies

Mass spectrometry is currently one of the most important techniques at the disposal of the structural chemist and it is not surprising that its application to the peptide problem is being intensively studied. Peptides are generally degraded in the mass spectrometer by the stepwise expulsion of amino acid residue units from the *C*-terminal end of the chain (e.g. **68** → **69** → **70**; **68** → **70**). Other patterns of degradation are found including side chain decomposition, but these variations are well documented and offer no barrier to the determination of residue sequences.

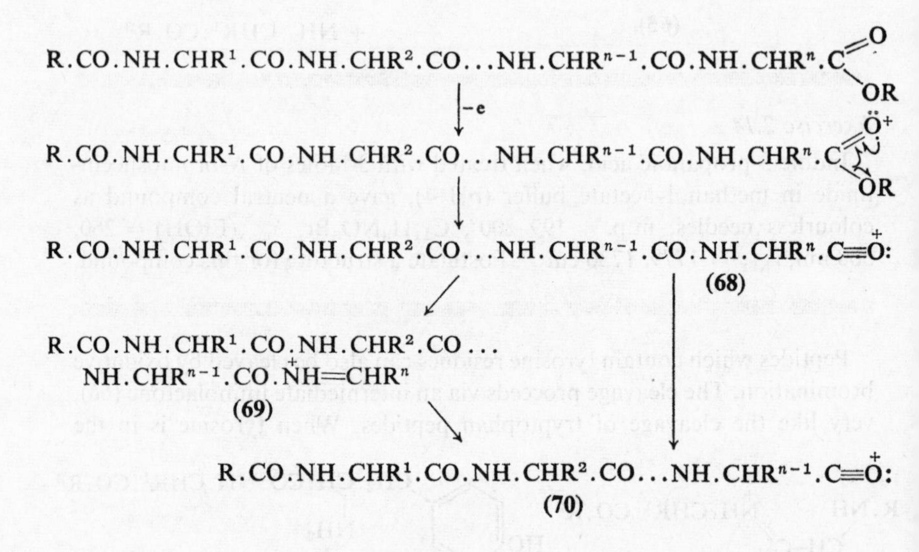

At present, the most severe limitation to the method is the necessity to convert peptides to volatile derivatives before committing them to the mass spectrometer. Two types of derivative have received the most attention. In the first, the terminal amino group of the peptide is acylated and the carboxyl group converted to an ester; in the second, the peptide amide groups throughout the chain are *N*-alkylated. The *N*-acyl ester derivatives are more volatile than peptides because they are not zwitterions; it is thought that the relative volatility of the *N*-alkylated derivatives stems from the absence of hydrogen bonding (Chapter 7).

Even these derivatives are only of moderate volatility and something like twelve residues in the molecule is the present upper limit to the method. However, peptide derivatives of workable size can be obtained by partial degradation of larger structures and, since the derivatives are relatively volatile, it should prove possible to purify them by vapour phase chromatography. It is therefore realistic to visualize an automated assembly, in which even proteins are degraded to small peptides, which, after purification, are 'sequenced' by on-line mass spectrometry. The process will presumably be computerized and the amino acid residue sequence of the protein will appear on a tape print-out. In theory, even an amino acid analysis could be rendered superfluous thereby but, in practice, it is likely that the most reliable results will be obtained if this and all other available chemical information are taken into account in the programme.*

Supplementary exercises

Exercise 2.15

Peptides have been methylated prior to mass spectrometry by treating them with methyliodide in the presence of silver oxide. *N*-Methylation is the normal course of events, but other reactions have been recorded. Provide reasonable explanations for the following observations:

(a) CH_3CO.Trp.Val.Glu.OMe methylates as expected with MeI/Ag_2O whereas CH_3CO.Glu.Glu.Ala.Glu.Ala.Tyr.Glu.OH gives a complex mixture of products which include

```
m/e                              OMe                 Me
                                  |                   |
746        └(Me)Glu.(Me)Ala.(Me)Glu.(Me)Ala.(Me)Tyr.(Me)Gly.OMe

                 OMe             OMe                 Me
                  |               |                   |
903 └(Me)Glu.(Me)Glu.(Me)Ala.(Me)Glu.(Me)Ala.(Me)Tyr.(Me)Gly.OMe

        OMe      OMe             OMe                 Me
         |        |               |                   |
977 (Me)Glu.(Me)Glu.(Me)Ala.(Me)Glu.(Me)Ala.(Me)Tyr.(Me)Gly.OMe
```

* Although the technique is outside the scope of this book, reference must be made to the success of x-ray crystallography in the determination of the structures of crystalline peptides. At maximum resolution this technique probably gives the most unequivocal results and, by its use, the full three dimensional structures of several proteins have been determined (Chapter 7).

(b) Acetyl methionine in the presence of MeI/Ag_2O is transformed into a volatile, crystalline cyclopropane derivative $C_8H_{13}NO_3$. Similarly

Glu.Gly.Pro.Trp.Met.OMe

is converted to a species with parent ion m/e 636.

Exercise 2.16

Acidic hydrolysis of Compound A (m.p. = 277–279, *M*(x-ray) = 769 ± 6), gave 1 mole of ammonium chloride plus the following amino acids in the molar proportions shown: valine (1), leucine (2), proline (1), phenylalanine (1), serine (1), aspartic acid (1). Compound A, although soluble in several organic solvents, was insoluble in water. It did not react with 2,4-dinitrofluorobenzene under the normal conditions. Four dipeptides, B–E, were isolated by partial acid hydrolysis of compound A; they had the following compositions: B, leucine (1), aspartic acid (1); C, valine (1), aspartic acid (1); D, serine (1), phenylalanine (1); E, leucine (1), aspartic acid (1). End-group determinations on peptides B–E gave DNP-aspartic acid, DNP-valine, DNP-serine and DNP-aspartic acid respectively. Mild hydrolysis of Compound A gave two larger peptides with the *N*-terminal sequences (Edman).

F,Leu.Pro.Val...
G,Phe.Leu.Pro...

Electrophoresis studies showed that:

(i) At pH 10, B and E both moved approximately twice as far as D towards the anode.
(ii) at pH 3, B moved further than E towards the cathode.

Explain these observations in terms of possible structures for *A*.

Exercise 2.17

Outline an approach which could be used to determine the arrangement of the disulphide bonds in a peptide which contained several cystine residues.

CHAPTER THREE

Chemical Synthesis I

General considerations

A structure which has been proposed on the basis of degradative studies is not usually accepted without reservations until it has been confirmed by synthesis. In the peptide field, and with proteins in particular, this ideal is difficult to realize, but peptide synthesis has reached a high degree of refinement. Many hundreds of peptides related to natural structures have been prepared and synthesis is often favoured instead of extraction for the commercial production of small peptides. At this moment, the synthesis of proteins seems a viable proposition.

One early approach to the synthesis of peptides employed α-chloroacyl chlorides (**71**) in the following manner (**71** → **72** → **73** etc.).

$$\underset{(\mathbf{71})}{\mathrm{Cl.CHR^{n-1}.CO.Cl}} + \mathrm{NH_2.CHR^{n}.CO_2^-} \xrightarrow{\mathrm{B}}$$

$$\underset{(\mathbf{72})}{\mathrm{Cl.CHR^{n-1}.CO.NH.CHR^{n}.CO_2^-}} + \overset{+}{\mathrm{B}}\mathrm{HCl^-} \xrightarrow{\mathrm{NH_3}}$$

$$\underset{(\mathbf{73})}{\overset{-}{\mathrm{Cl}}\overset{+}{\mathrm{N}}\mathrm{H_3.CHR^{n-1}.CO.NH.CHR^{n}.CO_2^-}} \xrightarrow[\mathrm{Cl.CHR^{n-2}.CO.Cl}]{\mathrm{2B}}$$

$$\mathrm{Cl.CHR^{n-2}.CO.NH.CHR^{n-1}.CO.NH.CHR^{n}.CO_2^-} + 2\overset{+}{\mathrm{B}}\mathrm{HCl^-} \longrightarrow \text{etc.}$$

*Exercise 3.1**

What inherent difficulties can you predict for this approach?

An octadecapeptide was prepared in this way, but the method suffers from several disadvantages, not the least of which are the inaccessibility of the starting materials and the difficulty of achieving stereochemical control. Routes based on the use of α-amino acids are superior. The optically pure α-amino acids can be obtained relatively easily from natural sources, or by synthesis, and the coupling together of these units to form peptides, unlike the build-up from the α-halogen derivatives, does not involve displacement reactions at the asymmetric α-carbon atoms.

It is not to be expected that α-amino acids will react together readily as zwitterions to give peptides, but little more than suppression of the zwitterion is necessary to make peptide synthesis thermodynamically feasible. This is illustrated by the formation of alanylglycine (**74**) from alanine (**75**), or alanine methyl ester (**76**), and glycine. Even the methyl ester is sufficiently reactive for aminolysis to occur.

$$\overset{+}{N}H_3.CHMe.CO_2^- + \overset{+}{N}H_3.CH_2.CO_2^- \longrightarrow$$
(**75**)

$$\overset{+}{N}H_3.CHMe.CO.NH.CH_2.CO_2^- + H_2O \ldots \Delta F = +4 \text{ kcal.}$$
(**74**)

$$NH_2.CHMe.CO_2Me + \overset{+}{N}H_3.CH_2.CO_2^- \longrightarrow$$
(**76**)

$$(\mathbf{74}) + MeOH \ldots \Delta F = -5 \text{ kcal.}$$

*Exercise 3.2**

Do you think that peptide synthesis could be accomplished satisfactorily from mixtures of α-amino acids and amino acid methyl esters as in the above equations?

Although the equilibrium in the reaction between methyl alaninate and glycine is on the side of the dipeptide, this reaction is not favoured kinetically. It is relatively slow and other reactions, especially formation of 2,5-dioxo-3,6-dimethylpiperazine (**77**) by self-condensation of alanine methyl ester, are more important.

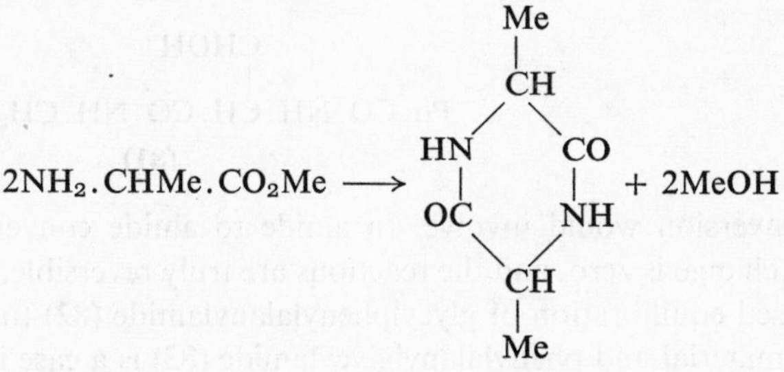

$$2NH_2.CHMe.CO_2Me \longrightarrow \text{(77)} + 2MeOH$$

(**77**): cyclo[HN–CH(Me)–CO–NH–CH(Me)–CO]

It is possible, by the use of more reactive derivatives than methyl esters to accelerate the coupling reaction, whilst undesirable side reactions can be avoided by the use of protecting groups. The modern approach to peptide synthesis has developed along these lines. However, before considering this development in detail, it is instructive to examine briefly an alternative approach which does not, at present, constitute an effective means of synthesising peptides, but which is of considerable theoretical interest and which almost certainly mirrors some biosynthetic processes.

In this approach to peptide synthesis, the two components of the coupling are sterically constrained so that the frequency of desired molecular collisions is increased, both in an absolute sense and relative to the frequency of undesirable collisions. In practice, the components are attached to a carrier molecule which affords the required stereochemistry, and the formation of the peptide bond is a matter of intramolecular rearrangement. The aminoacyl insertions (**78** → **79**; **80** → **81**) are reactions of this type.

It is interesting to note that the reverse reaction is not observed in these instances because of the unfavourable free energy change which an amide

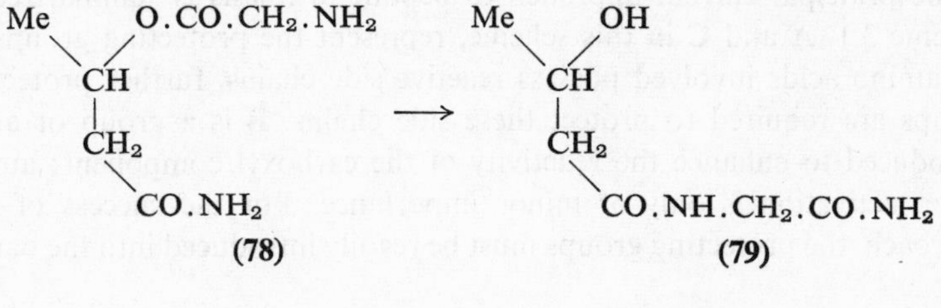

$$MeCH(O.CO.CH_2.NH_2).CH_2.CO.NH_2 \; \textbf{(78)} \longrightarrow MeCH(OH).CH_2.CO.NH.CH_2.CO.NH_2 \; \textbf{(79)}$$

$$\begin{array}{l} \quad\quad\quad\ \ \mathrm{Me} \\ \quad\quad\quad\ \ | \\ \quad\quad\quad\ \ \mathrm{CH.O.CO.CH_2.NH_2} \longrightarrow \\ \quad\quad\quad\ \ | \\ \mathrm{Ph.CO.NH.CH.CO.NH_2} \\ \quad\quad\quad\ \ \textbf{(80)} \end{array}$$

$$\begin{array}{l} \quad\quad\quad\quad\quad\ \mathrm{Me} \\ \quad\quad\quad\quad\quad\ | \\ \quad\quad\quad\quad\quad\ \mathrm{CHOH} \\ \quad\quad\quad\quad\quad\ | \\ \mathrm{Ph.CO.NH.CH.CO.NH.CH_2.CO.NH_2} \\ \quad\quad\quad\quad\quad\quad\quad\quad\quad \textbf{(81)} \end{array}$$

to ester conversion would involve. In amide to amide conversions, the free energy change is zero, and the reactions are truly reversible. The rapid base-catalysed equilibration of glycylphenylalanylamide (**82**) to a mixture of starting material and phenylalanylglycylamide (**83**) is a case in point.

$$\underset{\textbf{(82)}}{\mathrm{NH_2.CH_2.CO.NH.CH(CH_2Ph).CO.NH_2}} \xrightarrow[\text{anhydrous solvent, 30 minutes}]{\mathrm{Bu^tO^-}}$$

$$\textbf{(82)} + \underset{\textbf{(83)}}{\mathrm{NH_2.CH(CH_2.Ph).CO.NH_2.CH_2.CO.NH_2}}$$

One striking example of the potential effectiveness of these intramolecular reactions is provided by the spontaneous rearrangement of *O*-glycylsalicylic acid (**84**) to salicoylglycine (**85**). This reaction, which occurs with the participation of a free carboxyl group (!), has been employed in the synthesis of simple peptide derivatives (e.g. **86**, **87**).

Several factors militate against the synthesis of free peptides in this way. The preparation of the molecule which is expected to rearrange is often complex and may call for vigorous conditions: cleavage of the completed peptide from the carrier molecule is problematical; the general applicability of the rearrangement may be doubted since it is clearly conformation-dependent. However, this approach does avoid some of the pitfalls of the more usual procedure and it serves as a reminder of the reactivity of the peptide bond itself.

The principal current approach to peptide synthesis is summarized in Scheme 3.1. **A** and **C** in this scheme, represent the protecting groups. If the amino acids involved possess reactive side chains, further protecting groups are required to protect these side chains. **B** is a group or atom introduced to enhance the reactivity of the carboxyl component; amino group activation is only of minor importance. For the success of this approach, the protecting groups must be readily introduced into the parent

$O.CO.CH_2.NH_2$ / CO_2H (84) $\xrightarrow[\text{(ii) } CH_2N_2]{\text{(i) } H_2O \text{ rearrangement}}$ OH / $CO.NH.CH_2.CO_2Me$ (85) $\longrightarrow$

$O.CO.CH(CH_2Ph).NH_2$ / $CO.NH.CH_2.CO_2Me$ $\longrightarrow$

OH / $CO.NH.CH(CH_2Ph).CO.NH.CH_2.CO_2Me$ (86) $\dashrightarrow$

OH / $CO.Gly.Phe.Gly.NH_2$ (87)

amino acid moieties; they must prevent the group which they are blocking from interfering in the coupling reaction; and they must be easily removed at the conclusion of the synthesis, under conditions which will not damage the peptide. Carboxyl-activation and the subsequent coupling reaction must proceed without damage to the protecting groups and with a minimum of side-reactions. At all stages the dissymmetry of the amino acid moieties must be preserved.

Two ways of proceeding beyond the dipeptide stage are shown in Scheme 3.1. These require the selective removal of the protecting groups **A** or **C**, a sophistication which is not necessary for the preparation of the free dipeptide (route b). Chain extension is usually carried out from the *N*-terminal residue (route a), rather than from the *C*-terminal residue (route c). The reasons for this, which are connected with conserving the optical purity of the peptide, will be discussed later (p. 101). The coupling together of peptides, which is related to route *c*, must also be attempted circumspectly.

Modern peptide synthesis owes its success equally to the ingenious protecting and coupling methods which have been devised. The remainder of this chapter will be concerned with protecting groups; coupling methods will be dealt with in the next.

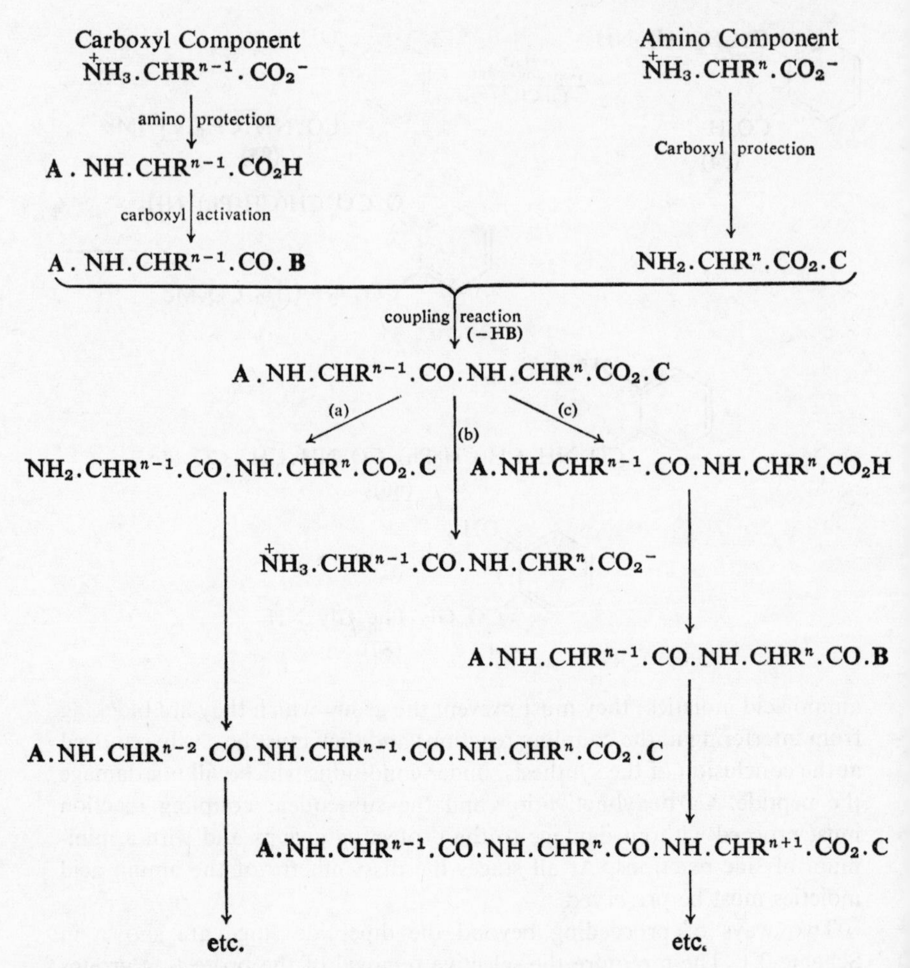

Scheme 3.1. Generalized route for the synthesis of peptides

Protecting groups

Terminal carboxyl group protection

Simple *N*-acyl derivatives of amino acids (**88**) can be prepared by reacting a salt of the amino acid with, for example, the corresponding acyl chloride. When an equimolar amount of base is slowly added during the course of the acylation, this is the well-known Schotten–Baumann reaction.

$$R.CO.Cl + NH_2.CHR.CO_2^-Na^+ + NaOH \longrightarrow R.CO.NH.CHR.CO_2^-Na^+ + NaCl + H_2O \quad \textbf{(88)}$$

The formation of the salt of the amino acid breaks the zwitterion. Carboxylate salts have also been used in peptide-forming reactions, but whereas they tend to be soluble in water, activated carboxyl components are generally soluble in organic solvents. Aqueous solvents are therefore employed in these coupling reactions and, under these conditions, various experimental difficulties are created and hydrolysis of the activated carboxyl component can compete with aminolysis. Covalently-bound carboxyl-protecting groups, which give amino components soluble in organic solvents, are therefore preferred and, in this context, esters are particularly important.

The hydrolysis of simple alkyl esters of amino acids and peptides under alkaline conditions is sufficiently faster than the comparable hydrolysis of the peptide bond for these derivatives to be useful as carboxyl-protected forms, i.e. amino components, during formation of the peptide bond. The sodium salt of the carboxylic acid, and hence the free acid, can usually be obtained in satisfactory yield by saponification of the final peptide ester. Because they are themselves susceptible to aminolysis, the amino acid esters must be stored in a form in which the nucleophilic properties of the amino group are suppressed. Generally, the esters are formed and stored as base hydrochlorides and the free amino groups are liberated immediately before the coupling reaction is carried out. Often the free base is liberated *in situ* by the addition of a tertiary base to the reaction mixture.

---

*Exercise 3.3**

Certain β-benzyl esters of aspartyl peptides (**89**) have been found to hydrolyse much more readily than benzyl esters of simple carboxylic acids (e.g. **90**). Devise a mechanism for the hydrolysis of the aspartyl esters which might account for this difference.

$$\begin{array}{c} CH_2.CO.O.CH_2.Ph \\ | \\ R.NH.CH.CO.NHR^1 \end{array} \qquad CH_3.CH_2.CO.O.CH_2.Ph$$

(89) **(90)**

It is an inconvenience to have to store amino acid esters as the amine salts and in some cases, undesirable side effects have been associated with the presence of tertiary base salts during the coupling reaction. Side reactions have also been noted during the subsequent saponification, even though this is usually accomplished with little more than one equivalent of alkali. Probably the most serious side reaction occurs at asparaginyl and aspartic acid residues where $\alpha \rightarrow \beta$ aspartyl rearrangements can occur (p. 14). Less frequent is the occurrence of base-catalysed racemization, either by direct ionization of the α-hydrogen atom, or, by an elimination process (p. 53, 99).

These side reactions could be avoided if the esters employed were more susceptible to alkaline hydrolysis, i.e. cleaved under even milder conditions, but then they would probably compete with the carboxyl component for the amine during the coupling stage. As the next chapter will show, one way to activate the carboxyl group is to convert it to an activated ester. It follows that improved carboxyl-protecting groups need to be cleaved by means other than saponification.

---

*Exercise 3.4**

So far, the acidic hydrolysis of esters has not been considered. Outline the mechanism of this type of reaction. Remember that, with simple alkyl esters, cleavage of the acyl–oxygen bond $R^1CO–OR$ occurs under both alkaline and acidic conditions. Some esters under acidic conditions are more likely to cleave with fission of the oxygen–alkyl bond $R^1CO.O–R$. What sort of esters might behave in this way and could they perhaps be employed advantageously as carboxyl protected forms in peptide synthesis?

Despite superficial similarities, the acidic hydrolysis of esters and saponification differ fundamentally. The acidic hydrolysis is reversible; saponification is not. However, with simple alkyl esters, both reactions involve cleavage of the acyl–oxygen bond ($R^1CO–OR$).

When the alcohol component is more complex, a second mechanism can become important. The significance of this mechanism, which involves cleavage of the oxygen–alkyl bond ($R^1CO.O–R$), is related directly to the stability of the carbonium ion (R^+). In some extreme cases, for example, triphenylmethyl (trityl) esters, this effect is so marked that partial ionization of the ester probably occurs even in neutral solution. Dependent on the solvent and on the character of the groups involved, the ions produced in this way **(91, 92)** can behave as independent entities or may stay 'loosely associated' in the form of ion-pairs.

$$R^1.CO_2R \rightleftharpoons \underset{\mathbf{(91)}}{R^1.CO_2^-} + \underset{\mathbf{(92)}}{R^+}$$

Such extreme behaviour is not necessary for the ester to possess a marked lability under acidic conditions. In these circumstances, an initial positive species, generated by protonation of the carbonyl group, will facilitate the dissociation of the carbonium ion. The change of mechanism with increasing complexity of the alkyl portion of the ester is strictly comparable to the change of mechanism observed in the solvolysis of alkyl halides of increasing complexity. Unimolecular reactions in general increase in importance in the series $Me < Et < Pr^i < PhCH_2 < Bu^t < Ph_2CH < Ph_3C$. This sequence holds good for halides, esters, ethers, alcohols, etc., and the reactions of these different classes of compounds only differ significantly in the point in the series at which the unimolecular mechanism becomes dominant.

Exercise 3.5

To what do you attribute the stability of tertiary butyl and triphenylmethyl carbonium ions?

t-Butyl esters **(93)** prove to be of about the right order of lability for peptide synthesis and these esters have been much used in recent years. It will be recalled that elimination vies with substitution in any reaction involving a carbonium ion intermediate. Even when cleaved in the presence of water, *t*-butyl esters form little if any butanol, so that isobutene is presumably the product. There is some evidence, however, that *t*-butyl chloride is formed when hydrochloric acid is used in the cleavage. The reaction can also be carried out in the complete absence of water since water does not play a part in the facile unimolecular cleavage, and trifluoroacetic acid, which is a strong acid and a good solvent, is a favoured medium for this reaction. Trifluoroacetic acid is readily evaporated after the cleavage, but apart from this manipulative advantage, the use of an anhydrous agent of this type also precludes unwanted hydrolysis of the peptide bond. Recently, it has been found that anhydrous hydrogen fluoride brings about the acidolytic cleavage of a wide range of protecting groups. The peptide bond is stable in this medium which will probably find wide application despite the hazards associated with its use.

The *t*-butyl carbonium ion has been known to create difficulties during cleavage due to its ability to react with amino acid side chains, in particular, with methionine, and tryptophan. In these cases, carbonium ion-scavengers, for example anisole or an excess of methionine, are sometimes added to the reaction mixture.

$$\underset{\textbf{(93)}}{R.\overset{\overset{\displaystyle O}{\|}}{C}.O.CMe_3} \xrightleftharpoons{H^+} R.\overset{\overset{\displaystyle \overset{+}{O}H}{\|}}{C}-O-CMe_3 \rightleftharpoons$$

$$R-C\begin{matrix}\nearrow OH \\ \searrow O\end{matrix} + \overset{+}{C}Me_3 \rightleftharpoons R.CO_2H + CH_2{=}CMe_2 + H^+$$

The unimolecular cleavage process is reversible, and ester synthesis may be achieved by careful control of the conditions. Amino acid *t*-butyl esters are prepared by reacting the amino acid with isobutene in the presence of catalytic amounts of sulphuric acid.

*Exercise 3.6**

Although glycine *t*-butyl ester (and, to some extent, alanine *t*-butyl ester) has been found to react appreciably with itself when refluxed for prolonged periods in an inert solvent, it is stable as the free base at room temperature. *t*-Butyl esters of other amino acids tend to be stable in the free base form even on heating. Account for this behaviour.

The formation of amine salts from amino acid *t*-butyl esters must be carried out with caution to ensure that the ester is not exposed to excess strong acid. Thus, the esters have been isolated as amine phosphites and aqueous citric acid is often employed instead of mineral acids to wash out the excess amino-component after a coupling reaction involving *t*-butyl esters. However, salt formation is not necessary to stabilize *t*-butyl esters and, unlike simple alkyl esters, the *t*-butyl esters can be stored in the free base form. Presumably, this stability is due to the steric effect of the *t*-butyl group which hinders approach of the nucleophilic amino group to the carbonyl, coupled with the strong $(+I)$ effect of the *t*-butyl group which reduces the reactivity of the ester carbonyl towards nucleophiles. Although a further σ-bond is involved, this inductive effect has also been said to enhance the nucleophilicity of the amino group.

Conceivably, trityl esters of amino acids could prove useful in peptide synthesis, but they are not easily prepared and would, perhaps, be too labile. Diphenylmethyl (benzhydryl) esters are readily accessible and find application. As may be expected, they are even more labile than *t*-butyl esters and are readily cleaved with 0·15–0·2M hydrogen chloride in an anhydrous solvent, for example, nitromethane. These esters can also be cleaved by catalytic hydrogenolysis, a technique not so far discussed, which leaves peptide bonds intact.

---

*Exercise 3.7**

What conditions do you anticipate might be used for the cleavage of (a) benzyl, (b) *p*-methoxybenzyl, (c) *p*-nitro-benzyl and (d) trimethylbenzyl esters?

Benzyl esters (**94**) are readily cleaved by hydrogenolysis at room temperature and atmospheric pressure over a catalyst composed, for example, of palladium adsorbed on finely divided charcoal, and consequently they are very useful for carboxyl protection. These esters are more resistant to acidic conditions than *t*-butyl or diphenylmethyl esters, but they are cleaved, although slowly, by treatment with a solution of hydrogen bromide in acetic acid. It will become apparent that this reagent is often used in peptide synthesis. The product of the cleavage is, presumably, benzyl bromide. In contrast, *p*-nitrobenzyl esters are not cleaved with hydrogen bromide in acetic acid, as the reagent is normally used, and this difference enables one to speculate in broad terms on the mechanism of the cleavage. It seems probable that protonation is the first step, followed by separation of the benzyl carbonium ion (route a) and possibly, displacement at the benzyl carbon atom (route b).

$$R{-}\overset{O}{\overset{\|}{C}}{-}O{-}CH_2Ph \underset{}{\overset{H^+}{\rightleftharpoons}} \left[R{-}\overset{^+OH}{\overset{\|}{C}}{-}O{-}CH_2Ph \right]$$

(94)

(a) $\rightleftharpoons$ $R{-}C(OH){=}O + PhCH_2^+$ $\xrightarrow{Br^-}$ $R.CO_2H + Ph.CH_2.Br$

(b) $\rightleftharpoons$ ^-Br + $R{-}C({=}\overset{+}{O}H){-}O{-}CH_2{-}Ph$ $\rightarrow$ $R.CO_2H + Ph.CH_2.Br$

The presence of a *p*-nitro substituent would tend to diminish the importance of route (a) since the *p*-nitrobenzylcarbonium ion would be expected to have a lower order of stability than the benzylcarbonium ion. Protonation might also be inhibited to some extent although the electronic effect of the nitro group would need to be transmitted through three σ-bonds. The stability of *p*-nitrobenzyl esters under these conditions makes one suspect that route (b) is relatively unimportant. As a corollary, *p*-methoxybenzyl esters should be more readily cleaved under acidic conditions than benzyl esters and this is confirmed in practice. *p*-Methoxybenzyl esters are readily cleaved by anhydrous trifluoroacetic acid.

Side-chain protection: carboxyl, hydroxyl and thiol groups

Carboxyl groups in side chains are sufficiently similar to α-carboxyl groups for the same protecting methods to be applicable to them. Methyl and ethyl esters are best avoided because of the dangers of rearrangement outlined above.

Syntheses can often be planned so that side-chain hydroxyl groups do not interfere in the coupling reaction, but in some cases, it is preferable to protect both aliphatic hydroxyl groups, as in serine and threonine, and phenolic groups, as in tyrosine; otherwise, ester formation, β-elimination or other undesirable side reactions might occur. The hydroxyl groups can usually be protected satisfactorily by converting them to ethers. Both *t*-butyl and benzyl ethers, cleaved by acidolysis and hydrogenolysis respectively, are valuable in this context.

Exercise 3.8

t-Butyl ethers can be prepared from amino- and carboxyl-protected tyrosine derivatives by reacting them with isobutene in the presence of an acidic catalyst. The corresponding ether of 3-nitrotyrosine cannot be prepared from an amino and carboxyl-protected form of 3-nitrotyrosine in this way. How might this difference be explained?

Exercise 3.9

The benzyl group is cleaved from *O*-benzyltyrosine by treatment with hydrogen bromide in acetic acid solution. *O*-Methyltyrosine is rather more resistant to this reagent. The methyl ether of 3-nitrotyrosine appears to be completely stable under these conditions; the corresponding benzyl ether is slowly cleaved. Postulate a mechanism for the cleavage reaction.

The protection of side-chain thiol groups during peptide synthesis is mandatory. Thiols, of course, are more acidic than hydroxyl compounds and have a pronounced nucleophilicity. Furthermore, it is not usually advisable to work with the thiols in the disulphide form, as in cystine. Disulphides (**95, 96, 97**) readily participate in displacement reactions and disulphide interchange is a real danger. Cystine is usually incorporated as cysteine derivatives which are deprotected at the end of the synthesis and oxidized to the disulphide form. In general, thiols like hydroxyls, are blocked by converting them to ethers.

$$2R^1.S.S.R^2 \rightleftharpoons R^1.S.S.R^1 + R^2.S.S.R^2$$

(95) **(96)** **(97)**

*Exercise 3.10**

Account for the following observations: treatment of *S-t*-butylcysteine-*t*-butyl ester with trifluoroacetic acid at room temperature gives *S-t*-butylcysteine; under the same conditions, *O-t*-butylserine *t*-butyl ester is converted to serine.

t-Butyl ethers (**98**), like the corresponding esters, are prepared by direct alkylation with an excess of isobutene, i.e. an acid-catalysed addition of the hydroxyl compound to the unsaturated bond occurs. The ethers, like the esters, are not conveniently prepared from *t*-butanol. *S-t*-Butyl thioethers (**99**) can be prepared by the acid-catalysed addition of the thiol to isobutene, but the thioether can also be prepared under acidic conditions from the thiol and *t*-butanol. The *t*-butyl thioether from cysteine, for example, is obtained in 70 per cent yield by the acid-catalysed reaction between cysteine and *t*-butanol at 80–100°C. This is an indication of the pronounced nucleophilicity of sulphur compounds, which is associated with a relatively low order of basicity. Elimination is usually secondary to displacement when sulphur is the nucleophile.

The *t*-butyl ethers (**98**) are readily cleaved under acidic conditions, presumably by a mechanism analogous to ester cleavage, in which primary protonation is succeeded by elimination of the *t*-butyl carbonium ion, which possibly forms isobutene by elimination of a β-hydrogen atom. *t*-Butyl thioethers (**99**) are much more stable under acidic conditions and *S-t*-butylcysteine hydrochloride, for example, may be recrystallized unchanged from 4M hydrochloric acid. At least two factors may be invoked to account for this difference: first, the low basicity of sulphur will mean that the thioethers tend to be less readily protonated than ethers; second, the marked nucleophilicity of sulphur might make the reverse reaction, in which the thiol reacts with the carbonium ion, so much more significant that the equilibrium is displaced to the left.

Fortunately, another method is available for cleaving thioethers which means that the *t*-butyl thioether can be employed in peptide synthesis despite its resistance to acid cleavage. In this method the thioether is converted to a mercaptide. A few minutes with boiling aqueous mercuric chloride is sufficient to cleave *S-t*-butylcysteine, for example, and free

$$\underset{(98)}{R{-}O{-}C(Me)_2{-}Me} \underset{}{\overset{H^+}{\rightleftharpoons}} R{-}\overset{H}{O^+}{-}CMe_3 \rightleftharpoons ROH + \overset{+}{C}Me_3 \rightleftharpoons H^+ + ROH + CH_2{=}CMe_2$$

$$\underset{(99)}{R{-}S{-}C(Me)_2{-}Me} \overset{H^+}{\longleftarrow} R{-}\overset{H}{S^+}{-}CMe_3 \rightleftharpoons RSH + \overset{+}{C}Me_3 \rightleftharpoons H^+ + RSH + CH_2{=}CMe_2$$

cysteine may be obtained from the insoluble mercaptide, which precipitates, by treating it with hydrogen sulphide.

The cleavage of benzyl thioethers is also different. Catalytic hydrogenolysis is not a satisfactory process because of the problem of catalyst poisoning. Reduction with sodium in liquid ammonia, on the other hand, works quite well and, if the ammonia is dry, little damage is usually sustained by the peptide chain. This method of reduction has been extremely useful in peptide synthesis and, until recently, *S*-benzylcysteine was the only *S*-protected derivative of cysteine to be widely used. *S*-Benzylcysteine is relatively stable under acidic conditions and is not cleaved, for example, with hydrogen bromide in acetic acid solution.

*Exercise 3.11**

What factors may be responsible for the degrees of acid lability exhibited by the following compounds:

Starting material	DPM–Cys	DMB–Cys	TMB–Cys	Tri–Cys
Thiol [a] (per cent)	9	12	14	100
Thiol [b] (per cent)	9	12	14	100

[a] measured after one minute, and
[b] after sixty minutes at room temperature in hydrogen bromide in acetic acid (0·25M solutions, 10 equivalents of HBr). DPM = diphenylmethyl (benzhydryl); DMB = 4,4′-dimethoxydiphenylmethyl; TMB = 3,3′, 4,4′-tetramethoxydiphenylmethyl; Tri = triphenylmethyl (trityl).

The ease of formation of carbonium ions relative to the benzyl carbonium ion may be measured in terms of the rates of solvolysis of the corresponding halides. These rates suggest the sequence: benzyl $<$ diphenylmethyl $<$ 2,4,6-trimethylbenzyl $<$ 2,4,6,2′,4′,6′-hexamethyldiphenylmethyl $<$ 4,4′-dimethoxydiphenylmethyl $<$ triphenylmethyl. This sequence could also be predicted by considering the resonance forms of the carbonium ions. More directly, and perhaps more dramatically, the relative stabilities of the carbonium ions are indicated by the equivalent conductivities of the halides. In liquid sulphur dioxide, for example, the values for the corresponding chlorides are 0·0013, 0·005, 0·013, 3·82, 4·84, 7·70.

S-Tritylcysteine is readily prepared by the action of the thiol on trityl chloride in dimethylformamide solution and the *S*-trityl bond in the resulting compound is easily and very rapidly cleaved by the action of hydrogen bromide in acetic acid. The benzhydryl and substituted benzhydryl derivatives can be made in a similar manner, but cleavage with the hydrogen bromide reagent is not complete and this is almost certainly due to its reversibility. Due to the powerful nucleophilicity of the thiol, the equilibrium of the cleavage reaction is presumably to the left and it can be displaced even further to the left in the presence of excess halide. For example, the cleavage of *S*-(2,2′-dimethyl-4,4′-dimethoxydiphenylmethyl)-cysteine in hydrogen bromide in acetic acid at room temperature is reduced from 14 per cent to 0·5 per cent if the reaction is carried out in the presence of five equivalents of the benzhydryl halide. Unlike *S*-tritylcysteine and the substituted *S*-benzhydrylcysteine derivatives, *S*-benzhydrylcysteine cannot be prepared conveniently from the thiol and the aralkyl chloride in dimethylformamide solution. Even after heating, the yield in this reaction barely exceeds 50 per cent. However, if the reaction is carried out in acidic solution, an almost quantitative yield of the thioether is obtained.

S-Tritylcysteine is not obtained when trityl chloride is employed in this way and the cleavage of *S*-tritylcysteine in very dilute solutions of hydrogen bromide in acetic acid, cannot be supressed even when ten equivalents of trityl chloride are added. It is not clear on the basis of these relatively simple arguments why the trityl derivatives seem to be qualitatively and not simply quantitatively different to the benzhydryl derivatives. Clearly, some factor(s) is not being taken into account. For example, steric hindrance, solvation or possibly ion-pair formation could be important in this case. It is reassuring, however, to note that the preparation of *S*-tritylcysteine from the carbinol and cysteine in trifluoroacetic acid solution has been reported.

*Exercise 3.12**

As an alternative to the thiol-protecting route, attempts have been made to convert non-sulphur containing peptides into cysteine derivatives by appropriate reactions. Thus, L-serine residues in peptides may be converted, by reacting the *O-p*-toluene sulphonate with thiobenzoate or thioacetate (R.CO.S⁻), to *S*-benzoyl and *S*-acetyl-L-cysteine derivatives respectively. When tritylthiocarbinol (sodium salt) is used for the displacement, the *S*-tritylcysteine derivatives which result are completely racemized at the cysteine α-carbon atom. Rationalize these results.

Serine-containing peptides (**100**), when treated with hydrogen bromide in acetic acid, are converted to the corresponding acetates (**101**). It is usually quite convenient to work with these derivatives when they arise, although saponification is ultimately required to obtain the free hydroxy compounds. However, because of the danger of β-elimination in this system, there is little merit in using the acetyl residue to protect the hydroxyl group of serine. Elimination is not likely to occur so readily with *S*-acyl cysteine derivatives and compounds of this type have been suggested for use in peptide synthesis. *S*-Acetyl- and *S*-benzoylcysteine, in particular, seem useful from this point of view. Thioesters, in general, are even more susceptible than *O*-esters to saponification and the protecting groups in these cases are therefore readily cleaved by treatment with dilute aqueous alkali.

$$\underset{\textbf{(100)}}{\mathrm{R.NH.\overset{\overset{\displaystyle CH_2OH}{|}}{C}H.CO.R^1}} \longrightarrow \underset{\textbf{(101)}}{\mathrm{R.NH.\overset{\overset{\displaystyle CH_2.O.CO.CH_3}{|}}{C}H.CO.R^1}}$$

Instead of employing *S*-acyl cysteine derivatives as building units during the synthesis of peptides, attempts have been made to convert the more accessible serine-containing peptides to cysteine peptides (e.g. **102** → **103** → **104**; R^3 = COPh or COMe). When the sodium salt of triphenylmethylthiocarbinol ($R^3 = CPh_3$) is used instead of thiobenzoate ($R^3 = COPh$) or thioacetate ($R^3 = COCH_3$) in this reaction, a racemic product is obtained, probably because the reaction proceeds by an elimination–addition mechanism (**103** → **105** → **104**). This type of elimination is always possible with suitable leaving groups and has been observed, for example, with *β*-chloroalanine, *O*-diphenylphosphorylserine, *S*-dinitrophenylcysteine and *S*-cyanocysteine derivatives.

$$\underset{\textbf{(102)}}{R^1.NH.\underset{|}{\overset{CH_2OH}{}}CH.CO.NH.R^2} \longrightarrow \underset{\textbf{(103)}}{R^1.NH.\overset{CH_2O.SO_2C_6H_4Me}{CH}.CO.NH.R^2}$$

$$\textbf{(103)} \xrightarrow{R^3S^-} \underset{\textbf{(104)}}{R^1.NH.\overset{CH_2.SR^3}{CH}.CO.NH.R^2}$$

$$\textbf{(103)} \xrightarrow{R^3S^-} \underset{\textbf{(105)}}{R^1.NH.\overset{CH_2}{\overset{\|}{C}}.CO.NH.R^2} \xrightarrow{R^3SH} \textbf{(104)}$$

S-Acyl cysteine derivatives (**106**), and indeed *O*-acyl serine derivatives, undergo another type of reaction (**106** → **107**) which might complicate their use in peptide synthesis. This is the reverse of the ester-forming reaction (N → O acyl shift) which was postulated to account for the susceptibility to hydrolysis of peptide bonds involving the amino group of serine. Fortunately, because sulphur is such a good nucleophile, this reaction is only likely to be important when the *α*-amino group of the cysteine is unsubstituted. This rearrangement has also been observed in a series of *S*-alkoxycarbonyl compounds, including *S*-ethoxycarbonyl, *S*-octyloxycarbonyl, *S*-cyclohexyloxycarbonyl, and *S*-phenethoxycarbonyl cysteine derivatives. It is not observed with carbamoyl derivatives (e.g. **106**; R^2 = Et.NH.) and this may be attributed to the $(+M)$ deactivating effect of the carbamoyl nitrogen. Accordingly, *S*-ethylcarbamoylcysteine derivatives, which are cleaved very rapidly under mildly alkaline conditions but which are stable under acidic conditions, are potentially useful in peptide synthesis.

One sulphur-protecting group, which is particularly valuable because it is not removed under the conditions normally employed for cleaving other protecting groups in peptide synthesis, is the acetamidomethyl group.

$$R^2{-}\overset{\displaystyle O}{\overset{\|}{C}}{-}S{-}CH_2{-}CH(NH_2){-}CO.R^1 \;\; (HO^-) \rightleftharpoons R^2{-}\overset{\displaystyle O^-}{\overset{|}{C}}(S{-}CH_2)(HN{-}CH.CO.R^1) \rightleftharpoons R^2.CO.NH.\overset{\overset{\displaystyle {}^-S}{|}}{\overset{CH_2}{|}}{CH}.CO.R^1$$

(106) (107)

S-Acetamidomethylcysteine hydrochloride (**108**) is prepared by reacting cysteine with acetamidomethanol (**109**) in hydrochloric acid solution at pH 0·5 and 25°C. The protecting group is cleaved quantitatively by treatment with two equivalents of a mercuric salt at pH 4 and room temperature; it is unaffected by trifluoroacetic acid, by hydrofluoric acid at 0°C, by hydrogen bromide in ethanol, and by aqueous solutions over the pH range 0–13 at 25°C.

$$\overset{-}{Cl}\overset{+}{N}H_3.\overset{\overset{\displaystyle CH_2SH}{|}}{CH}.CO_2H + \underset{(109)}{CH_3.CO.NH.CH_2OH} \longrightarrow \underset{(108)}{\overset{-}{Cl}\overset{+}{N}H_3.\overset{\overset{\displaystyle CH_2.S.CH_2.NH.CO.CH_3}{|}}{CH}.CO_2H}$$

Exercise 3.13

Cysteine methyl ester hydrochloride reacted under reflux in methylene chloride with 2,3-dihydropyran to yield an oily substance which could not be solidified, but which gave, by saponification and subsequent neutralization, a crystalline solid $C_8H_{15}O_3NS$, m.p. = 186–187°C, $[\alpha]_D^{25}$ = +11·6°C (in water). Cysteine could be obtained from this compound by dilute acid hydrolysis or by treating it with aqueous silver nitrate. Outline the series of reactions involved. The compound has been proposed as an *S*-protected cysteine derivative which might be useful in peptide synthesis, but there is one particular drawback to its use which is inherent in the structure of the compound. Can you see the difficulty?

Exercise 3.14

Compare the formation and hydrolysis of acetals and ketals with the formation and hydrolysis of the sulphur analogues. Ketals have been prepared to protect hydroxyl groups during the synthesis of nucleotides (Chapter 8) by reacting the hydroxy component with 4-methoxy-5,6-dihydro-2H-pyran. What type of substituent in the carbonyl moiety will tend to make these derivatives more stable to acidic hydrolysis?

Exercise 3.15

Cysteine hydrochloride reacts with boiling anhydrous acetone to give a derivative (a thiazolidine) by a reaction which resembles ketal formation. The nitrogen of this derivative is appreciably nucleophilic and reacts, for example, with an acetic anhydride–formic acid mixture to give the crystalline *N*-formyl derivative, $C_7H_{11}NO_2S$. This compound, which is effectively an *N*,*S*-protected cysteine derivative, can be used in peptide synthesis and the protecting group can be readily cleaved afterwards. Consider the nature of these reactions.

Terminal amino group protection

In principle, the reactions of hydroxyl, amino, and thiol groups have much in common and derivatives characteristic of these groups can often be made in similar ways. Thus, *N*-benzyl (**110**; $R^1 = R^2 = H$, $R^3 = Ph$), *N*,*N*-dibenzyl (**111**), *N*-benzhydryl (**110**; $R^1 = H$, $R^2 = R^3 = Ph$) and *N*-trityl (**110**; $R^1 = R^2 = R^3 = Ph$) amino acids can be made via displacement reactions from the corresponding aralkyl halides in much the same way as ethers and thioethers. In turn, the conditions under which these aralkyl amines are cleaved resemble the conditions for the cleavage of thioethers. However, the basicity of the amino group is an additional complication.

$$R^1R^2R^3C.NH.CHR.CO_2H \quad \textbf{(110)}$$

$$(Ph.CH_2)_2N.CHR.CO_2H \quad \textbf{(111)}$$

The basicity of alkyl amines depends on factors which are not easy to assess. In aqueous solution, tertiary amines are usually less basic than the corresponding primary amines, which are less basic than the secondary amines. The series $NH_3 < RNH_2 < R_2NH$ may be explained in terms of the (+*I*) effect of the alkyl substituent, R, which will tend to increase electron availability on the nitrogen atom and hence, increase the basicity of the amine. On the other hand, the relative basicity of the tertiary amine obviously cannot be explained in terms of the inductive effect and other factors need to be considered.

If the basicity of the amines is measured in chlorobenzene, the amines, in order of increasing basicity, fall into the sequence which considerations of the inductive effect would predict, viz. $NH_3 < RNH_2 < R_2NH < R_3N$. It is argued, therefore, that the cations derived from primary and secondary amines in aqueous solution are stabilized by hydration, thus enhancing the basicity of the amine; whereas with tertiary amines, hydration is not so important because of steric hindrance. Other arguments have also been presented which consider that distortion of the pyramidal structure of the tertiary nitrogen, due to steric strain, exercises an electronic effect which would also be expected to reduce basicity.

*Exercise 3.16**

Relate the pH of a conjugate acid at half neutralization to its pK_a value (see appendix) and discuss the implications of the following data:

Amino acid derivative	pH at half neutralization[a]	Infrared (mull)
Glycine	9·90	No absorption in 1700–1750 cm^{-1} region.
N-DMB-glycine	8·20	
N-MMB-glycine	8·25	
N-benzyl-glycine	9·22	
N,*N*-dibenzyl-glycine	7·89	
N-trityl-glycine	6·40	

[a] The amino acid derivative was dissolved in 1:1 water: dioxan and a half equivalent of sodium hydroxide was added to the resulting solution. Abbreviations: MMB = 2,2′-dimethyl-4,4′-dimethoxybenzhydryl; DMB as in Exercise 3.11*.

*Exercise 3.17**

N-Tritylglycine methyl ester is readily saponified in aqueous acetone and the resulting solution when carefully acidified yields *N*-tritylglycine. The hydrazide of *N*-tritylglycine may be prepared by the action of hydrazine on the methyl ester. Neither of these reactions gives any appreciable amount of product in the case of *N*-tritylphenylalanine methyl ester, and the ester can be recovered from the reaction mixtures unchanged. Account for these observations.

The complexity of the problem is brought out by the consideration of a series of *N*-aralkylamino acids. In Exercise 3.16*, the pH values at half neutralization indicate the pK_a values of the various substituted amines (see appendix). Inspection reveals that the strengths of the amines diminish in the series glycine > *N*-benzylglycine > *N*-DMB-glycine > *N*-MMB-glycine > *N,N*-dibenzylglycine > *N*-tritylglycine. Electronic factors alone cannot account for the drop is basicity along this series and some other considerations, possibly steric, need to be taken into account. The strengths of the carboxylic acids increase in approximately the same order, *N*-tritylglycine being the strongest acid of the aralkyl glycines. This sequence can be explained in terms of the electronic effect of the substituted amino groups—the amine least willing to donate its electrons, i.e. the amine which possesses the relatively greater affinity for electrons, produces the strongest acid—but this reasoning is not adequate if solvation of the cation is invoked to account for the basicities of the amines. In this case the electron-donating tendency of the nitrogen atom is not the major factor which determines basicity and measurement of basicity will not necessarily indicate the true electron affinity of the nitrogen atom. Whatever its origins, the basicity exhibited by an aralkylamino acid is extremely important so far as its suitability for use in peptide synthesis is concerned.

The distinction between basicity and nucleophilicity is well illustrated by these compounds. Even *N*-tritylglycine is weakly basic and may be converted into a hydrochloride by treatment with one equivalent of hydrogen chloride in dioxan solution, but it is difficult to conceive that a tritylamino group could act as a nucleophile. The trityl group is very bulky and any reaction dependent on the nitrogen acting as a nucleophilic centre would be subject to severe steric hindrance. Indeed, the hindrance extends beyond the amino group and *N*-tritylamino acid esters, for example, are in general quite resistant to saponification.

DMB-amino acid esters, though not as hindered as their trityl counterparts, are still appreciably hindered and can only be saponified with difficulty. In consequence, the amino group in these derivatives is not likely to possess nucleophilic properties and, from this point of view, the protection of the amino group might be accounted satisfactory. On the other hand, the DMB-amino group is basic and this factor makes these derivatives less than favourable for use in peptide synthesis. The DMB-amino acids are finely crystalline compounds, generally insoluble in organic solvents other than hot ethanol and acetone, and they possess high, indefinite melting points. These properties suggest that the derivatives are zwitterions, a conclusion which finds further support in the infrared spectra of the compounds. Thus, DMB-glycine has no carbonyl absorption in the 1710–1750 cm^{-1} range, but, like an *N*-tritylamino acid, it can be converted to a hydrochloride which absorbs strongly at $\sim$1750 cm^{-1}.

---

*Exercise 3.18**

Aralkylamino acids are cleaved to give the free amino acids by heating them at 100°C in 50 per cent aqueous acetic acid for five minutes, to the following extents:

N-tritylamino acids	100 per cent
N-DMB-amino acids	100 per cent
N-MMB-amino acids	100 per cent
N-benzylamino acids	0 per cent
N,*N*-dibenzylamino acids	0 per cent

How can these differences be explained?

N-Benzyl, *N*,*N*-dibenzyl and the various substituted *N*-benzhydrylamino acids are very similar in the basicity of their amino groups. However, when acid lability is considered, the substituted benzhydryl derivatives are rather like *N*-tritylamino acids and totally unlike the *N*-benzyl and *N*,*N*-dibenzyl derivatives. The free amino acids are liberated rapidly from

N-trityl, *N*-DMB, *N*-MMB, and *N*-TMB-amino acids by warming them briefly in 50 per cent aqueous acetic acid. *N*-Benzyl and *N,N*-dibenzyl-amino acids are stable under these conditions and reduction with sodium in liquid ammonia is necessary to effect their cleavage. Once again, acid lability is clearly related to the stability of the various carbonium ions generated in the cleavage reactions.

The trityl group, because of its favourable protecting properties (reduction of basicity and elimination of nucleophilicity) and because it can be cleaved so readily, is useful as an amino-protecting group in peptide synthesis, but because of steric hindrance, the synthesis of *N*-tritylamino acids is not without difficulties and the range of methods which can be used at the coupling stage is limited. Other amino-protecting groups which do not suffer from these disadvantages are therefore superior.

---

*Exercise 3.19**

Early attempts to develop amino-protecting groups for use in peptide synthesis led to the investigation of *N*-ethoxycarbonylamino acids ($EtOCONHCHRCO_2H$). It was anticipated that the carbamate would be readily saponified from the protected peptide, without disturbing the peptide bonds, and that the carbamic acid produced by acidification would readily decarboxylate to the free amine. Unfortunately, peptide hydrolysis and other side reactions occurred during the alkaline treatment and the protecting group was not very satisfactory. These difficulties have been overcome and urethane groups of this general type constitute the most important protecting groups in use in peptide synthesis. Try to design groups of this type which should be cleavable selectively under conditions which will not effect the peptide bond.

One way to reduce the basic and nucleophilic properties of an amine is to convert it to an acyl derivative. If this leads to the production of a simple amide (**112**; R = alkyl), it is clear that selective cleavage of the acyl group in the presence of the peptide bonds will not be feasible. Hence, acetyl protection is impractical because the CH_3CO–NHR bond is only hydrolysed under conditions which would hydrolyse peptide bonds. It will become apparent in Chapter 4 that such protecting groups would also create racemization difficulties. Two simple acyl derivatives do find application but these are special cases. Trifluoroacetyl derivatives (**112**; R = CF_3) are very susceptible to hydrolysis because of the strongly electronegative substituents (the fluorine atoms) which the group bears, and the derivatives may be cleaved under alkaline conditions comparable to those used for the saponification of methyl and ethyl esters. Formyl derivatives (**112**; R = H) may be cleaved by acid catalysed methanolysis, or by oxidation with hydrogen peroxide in trifluoroacetic acid. Presumably the oxidation proceeds to the carbamate which undergoes decarboxylation to the free amine.

$$R.CO.NH.R^1$$
(112)

The cleavage of the formyl group via the carbamate provides a pointer to acyl-protecting groups which are of the greatest possible value. These are the alkoxycarbonyl and aralkoxycarbonyl (i.e. urethane-type) groups, which constitute carbamic ester derivatives (**112**; R = O.alkyl or R = O.aralkyl). Foremost in this category is the benzyloxycarbonyl (also called carbobenzoxy) group (**112**; R = OCH_2Ph). This was the first α-amino-protecting group to be applied really successfully in peptide synthesis and its importance is undiminished despite the development of several other excellent protecting groups of this type. It can be introduced by the usual Schotten–Baumann reaction from the acylchloride(benzylchloroformate) and the amino acid salt.

Exercise 3.20

Outline the reaction sequence involved in the preparation of an *N*-benzyloxycarbonylamino acid, starting from benzyl alcohol.

The *N*-benzyloxycarbonyl group may be cleaved by catalytic hydrogenolysis, for example, over palladized charcoal at room temperature and atmospheric pressure. Presumably toluene and the carbamic acid are formed and decarboxylation of the latter gives the free amine. Addition of acid is often helpful but not essential. Chemical reduction, for example, with sodium in liquid ammonia, can also be used for the cleavage. Alternatively, the benzyloxycarbonyl group can be cleaved by acidolysis, for example, by treatment with hydrogen bromide in acetic acid solution at room temperature. The type of mechanism which prevails during the acidolytic cleavage will depend on the conditions used.

N-p-Nitrobenzyloxycarbonylglycine ethyl ester (**113**; R = NO_2) cleaves 2400 × more slowly than the unsubstituted benzyloxycarbonyl derivative (**113**; R = H) in a sulphuric-acetic acid mixture. With hydrogen bromide in acetic acid, the *p*-nitro derivative cleaves only 13 × more slowly than the unsubstituted benzyloxycarbonyl compound. Similar effects are observed in other nucleophilic displacements involving benzyl compounds. They suggest that, after initial protonation, the sulphuric–acetic acid cleavage proceeds predominantly by carbonium ion formation, (route a), whereas the hydrogen bromide in acetic acid cleavage is mainly a bimolecular displacement reaction (route b). Again, ion pairs rather than the free ions may be involved in these reactions.

R–C_6H_4–$CH_2.O.CO.NH.CH_2.CO_2.Et$
(**113**)

H^+ ⇌

R–C_6H_4–$CH_2.\overbrace{O.CO.NH}^{H^+}.CH_2.CO_2.Et$

(a) ⇌ R–C_6H_4–$\overset{+}{C}H_2$ + $O{=}C(OH).NH.CH_2.CO_2.Et$

(b) X^- ⇌ $\left[\text{R–}C_6H_4\text{–}CH_2 \cdots \overbrace{O.CO.NH}^{H^{\delta+}}.CH_2.CO_2.Et \;\; \cdots X^{\delta-} \right]$

(+X^-) ⟶ R–C_6H_4–CH_2X + CO_2 + $NH_2.CH_2.CO_2.Et$

A study of the effect of various ring substituents, R, in the aniline moiety of benzyl carbanilates (**114**) gives an indication of the site of the initial protonation (*a priori*, protonation could involve either the oxygen or the nitrogen atoms of the urethane grouping). Pseudo-first order rate constants for the cleavage of the benzyl carbanilates under identical conditions are:

R	MeO	Me	Et	H	Cl	Br
$k \times 10^{-5}$ (sec^{-1})	15.6	11.7	12.0	9.2	7.1	6.7

It seems probable that these differences will only be concerned with the initial protonation since the group, R, is far removed from the displacement centre.

$$\text{Ph.CH}_2\text{.O.CO.NH}-\langle\text{C}_6\text{H}_4\rangle-\text{R}$$

(114)

A plot of $\log k/k_{\mathrm{H}}$ against the corresponding Hammett substituent constants (see Exercise 3.21* and appendix) for the carbanilate cleavages gives an approximately straight line of slope (i.e. ρ) ≈ -0.7. For the protonation of substituted anilines (**115**) the corresponding value is $\rho \approx -2{\cdot}7$. By comparison, the protonation of the carbanilate is relatively insensitive to the electronic effect of the substituent. Although admittedly not unequivocal, this evidence suggests that protonation involves the oxygen and not the nitrogen of the urethane grouping because the oxygen atom is further removed from the influence of the ring substituent.

$$\text{R}-\langle\text{C}_6\text{H}_4\rangle-\text{NH}_2 + \text{H}^+ \rightleftharpoons \text{R}-\langle\text{C}_6\text{H}_4\rangle-\overset{+}{\text{N}}\text{H}_3$$

(115)

---

*Exercise 3.21**

Cleavage of substituted benzyloxycarbonylglycines (**116**) with hydrogen bromide in acetic acid (0·85M HBr) at 25°C is first order up to 75 per cent completion. The pseudo-first order rate constants are:

R	H	*p*-MeO	*p*-Me	*p*-F	*m*-Cl
$k \times 10^{-5}$ (sec^{-1})	18	1500	160	14	5·5
σ	0·000	−0·268	−0·170	+0·062	+0·373

R	*p*-Cl	*p*-NO
$k \times 10^{-5}$ (sec^{-1})	9·7	2·6
σ	+0·227	+0·778

Determine the reaction constant(s) ρ involved (see appendix). What mechanism(s) can you propose for these cleavage reactions?

R-C_6H_4-$CH_2.O.CO.NH.CH_2.CO_2H$

(116)

The cleavage of benzyloxycarbonylamino acids and peptides is relatively independent of the nature of the amine but it is extremely dependent on the nature of substituents in the benzyloxycarbonyl group. This was demonstrated above in the case of the *p*-nitro derivative; information concerning other substituents is given in Exercise 3.21*. A plot of $\log k/k_H$ against the σ constants of the different substituents in these cases is very informative. The *p*-methoxy, *p*-methyl and unsubstituted benzyloxycarbonyl derivatives (negative σ values) lie on a straight line of $\rho \approx -7$; all of the other derivatives (positive σ values) lie on a straight line of $\rho \approx -1$. The unsubstituted benzyloxycarbonyl derivative lies approximately at the intersection of the two lines (see Figure 3.1).

This evidence strongly suggests that two mechanisms are involved. In one case ($\rho = -7$) a unimolecular process, in which a carbonium ion is generated from the protonated urethane derivative, is probably the predominant cleavage route; in the other ($\rho = -1$), attack by bromide ion on the protonated species is probably important. Cleavage of the acyl–O bond seems unlikely, a conclusion which is substantiated by the nature of the products. Where these have been fully investigated, virtually all of the benzyl moiety has been recovered as benzyl bromide.

The position of the unsubstituted benzyloxycarbonyl derivative at the intersection of the two graphs suggests that both unimolecular and bimolecular processes may be important during the cleavage with hydrogen bromide in acetic acid. Benzyloxycarbonyl derivatives can also be cleaved by refluxing them in trifluoroacetic acid. In this case, anisole, catechol or indole are usually added to the reaction mixture to react with carbonium

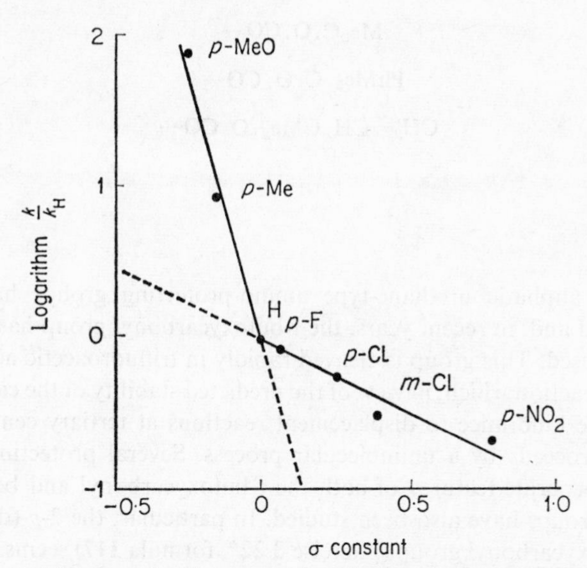

Figure 3.1. Cleavage of substituted benzyloxycarbonylglycines (see Exercise 3.21*). Redrawn from the data of K. Bláha and J. Rudinger, *Collection of Czechoslovak Chemical Communications*, (1965) **30**, 585.

ion products and thus avert side reactions. The introduction of electron-donating substituents into the benzyl moiety, i.e. carbonium ion-stabilizing substituents, will obviously increase the acid lability of these groups. The *p*-methoxybenzyloxycarbonyl group has proved very useful from this point of view, since it can be cleaved with trifluoroacetic acid at room temperature.

Exercise 3.22

Arrange the following alkoxycarbonyl and aralkoxycarbonyl groups in order of increasing acid lability.

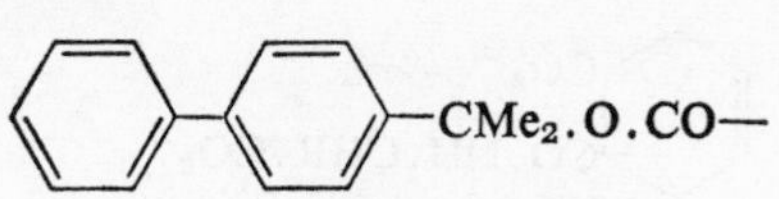

—CMe_2.O.CO—

(117)

$$Me_3C.O.CO—$$

$$PhMe_2.C.O.CO—$$

$$CH_2{=}CH.CMe_2.O.CO—$$

Various aliphatic urethane-type amino-protecting groups have been investigated and, in recent years, the *t*-butoxycarbonyl group has come to be widely used. This group is cleaved rapidly in trifluoroacetic acid in the cold by a reaction which, in view of the predicted stability of the carbonium ion and the hindrance to displacement reactions at tertiary centres, presumably proceeds by a unimolecular process. Several protecting groups which incorporate features of both the *t*-butoxycarbonyl and benzyloxycarbonyl groups have also been studied. In particular, the 2-*p*-(diphenyl)-isopropyloxycarbonyl group (Exercise 3.22*, formula **117**) seems valuable. It is cleaved quantitatively by 80 per cent acetic acid at 22–25°C, approximately three thousand times faster than the *t*-butoxycarbonyl group.

There are three other amino-protecting groups which merit special notice: the phthalyl, the *p*-toluene sulphonyl and the *ortho*-nitrophenylsulphenyl groups. The phthalyl group, as used in the Gabriel synthesis of amines, is useful because phthalyl derivatives can be cleaved by hydrazinolysis; alkaline hydrolysis is inappropriate because peptide bonds are also hydrolysed under these conditions, but formation of phthalhydrazide (**118**) by the action of hydrazine does not effect the peptide bond. Phthalamides (**119**), which are not cleaved by hydrazine, are produced when phthalyl derivatives are exposed to alkali, and therefore the use of alkaline conditions at any stage of peptide synthesis has to be avoided when this protecting group is present.

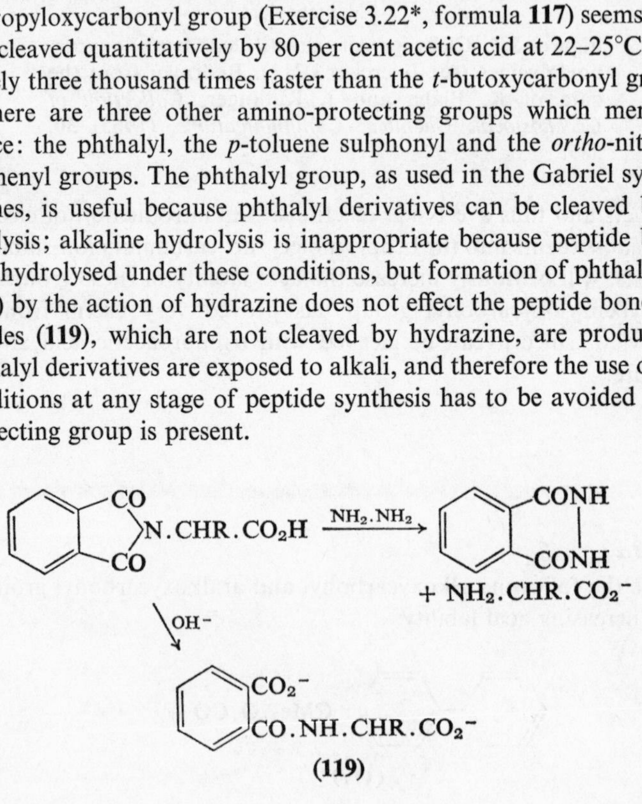

$$C_6H_4(CO)_2N.CHR.CO_2H \xrightarrow{NH_2.NH_2} C_6H_4(CONH)_2 \ (\mathbf{118}) + NH_2.CHR.CO_2^-$$

$$\downarrow OH^-$$

$$C_6H_4(CO_2^-)(CO.NH.CHR.CO_2^-)$$

(119)

N-p-Toluenesulphonyl (tosyl-) derivatives (**120**) are useful in peptide synthesis. They are prepared from the amine and tosyl chloride under Schotten–Baumann conditions. The tosyl-N bond is cleaved by reduction with sodium in liquid ammonia. *N-ortho*-nitrophenylsulphenyl derivatives (**121**), prepared from the amine and *ortho*-nitrophenylsulphenyl chloride are also valuable. They are cleaved rapidly when treated with two equivalents of hydrogen chloride in an inert solvent by a reaction which probably involves a direct displacement of the amine moiety from the sulphur atom by the halide.

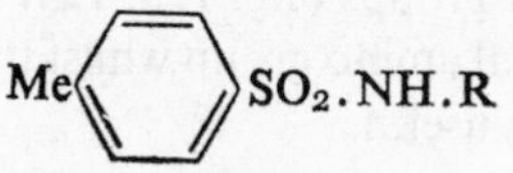

Me–C_6H_4–SO_2.NH.R (**120**)

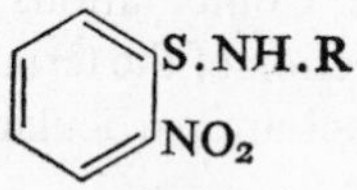

C_6H_4(S.NH.R)(NO_2) (**121**)

Exercise 3.23

Phthalylamino acids can be prepared by reacting amino acids with *N*-carboethoxyphthalimide. Outline the reactions involved, starting from phthalimide. What disadvantage would there be in preparing the *N*-phthalyl derivative by the Gabriel method, i.e. from potassium phthalimide and the alkyl halide?

Side-chain protection: guanidino, amino and imidazole groups

Three commonly-occurring amino acids possess basic side chains: histidine contains an imidazole moiety; lysine an ϵ-amino group; arginine a δ-guanidino group. The histidine side-chain is sometimes protected as the N^{im}-benzyl (im = imidazole) derivative (**122**) which can be cleaved by reduction with sodium in liquid ammonia; at other times it is not protected. Lysine and arginine side-chains must always be protected if they are to be prevented from participating in coupling reactions.

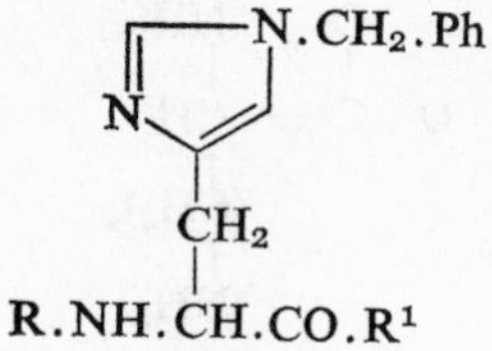

(imidazole ring with N.CH_2.Ph)–CH_2

R.NH.CH.CO.R^1

(**122**)

Exercise 3.24

Devise a synthesis of N^{α}-benzyloxycarbonyl-N^{im}-benzylhistidine.

In the case of lysine, all of the protecting groups discussed above for the protection of terminal amino groups, are also available for the protection of the side-chain. Combinations of groups (e.g. **123**, **124**) which allow the selective deprotection of the terminal amino group whilst the N^{ϵ}-protecting group is left intact are particularly useful.

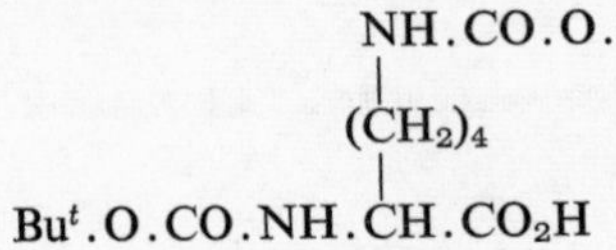

$$\begin{array}{c} \text{NH.CO.O.CH}_2\text{.Ph} \\ | \\ (\text{CH}_2)_4 \\ | \\ \text{Bu}^t\text{.O.CO.NH.CH.CO}_2\text{H} \end{array}$$

(123)

$$\begin{array}{c} \text{NH.CO.O.Bu}^t \\ | \\ (\text{CH}_2)_4 \\ | \\ \text{Ph.CH}_2\text{.O.CO.NH.CH.CO}_2\text{H} \end{array}$$

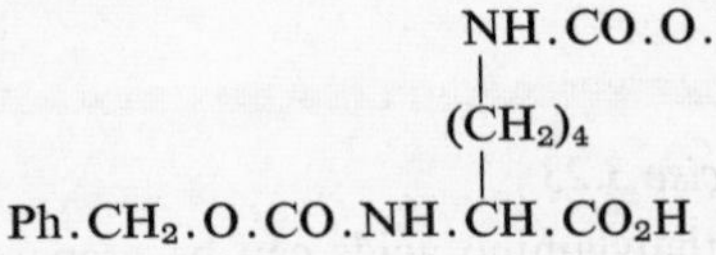

(124)

Selective reaction of the ϵ-amino group during the protecting stage can be achieved by using the copper complex of the amino acid (**125**), and subsequently removing the copper, for example, by reduction with H_2S or by the use of a sequestering agent. This method is of general applicability for blocking α-amino and α-carboxyl groups during the introduction of functions into the side chains.

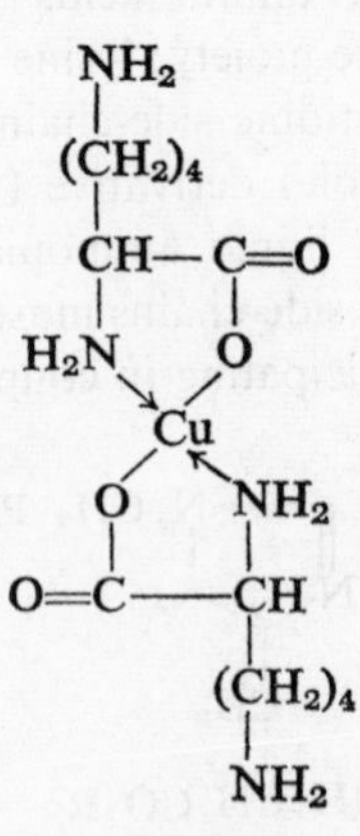

$$\begin{array}{c} \text{NH}_2 \\ | \\ (\text{CH}_2)_4 \\ | \\ \text{CH—C=O} \\ | \quad\quad | \\ \text{H}_2\text{N} \quad \text{O} \\ \searrow \;\; / \\ \text{Cu} \\ / \;\; \nwarrow \\ \text{O} \quad \text{NH}_2 \\ | \quad\quad | \\ \text{O=C——CH} \\ | \\ (\text{CH}_2)_4 \\ | \\ \text{NH}_2 \end{array}$$

(125)

The guanidino group of arginine is distinctive because of its pronounced basicity (p$K_a \approx 12 \cdot 5$) which is such that the group remains completely protonated under conditions in which the α-amino group (p$K \approx 9 \cdot 25$) is available to act as a nucleophile. Thus, the side-chain of the arginine residue can be protected as a salt, for example, as the hydrobromide, and used in peptide synthesis at pH 9 without the complication of side-reactions. The guanidino group is nucleophilic and can also be blocked by covalently-bound groups of the amine-protecting type. N^{α}-Benzyloxycarbonyl-N^{δ}-tosylarginine (**126**), in particular, has been used in this way.

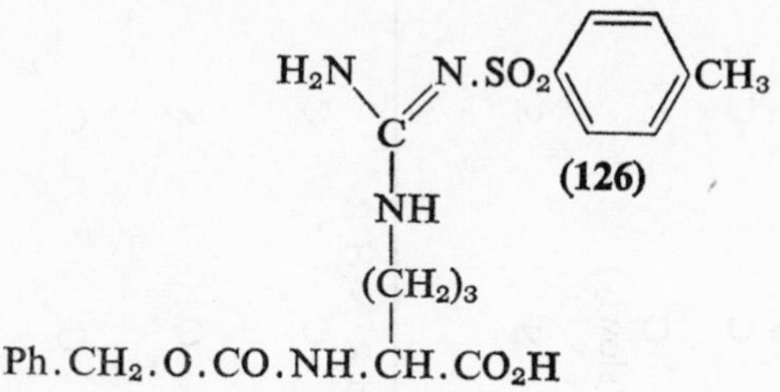

$$\begin{array}{c} H_2N \quad N.SO_2C_6H_4CH_3 \\ \diagdown \,\, /\!\!/ \\ C \\ | \\ NH \\ | \\ (CH_2)_3 \\ | \\ Ph.CH_2.O.CO.NH.CH.CO_2H \end{array}$$

(126)

Exercise 3.25

Outline a possible route for the synthesis of N^{α}-benzyloxycarbonyl-N^{δ}-tosylarginine (**126**).

One of the most popular methods of blocking the guanidino group is to nitrate it and many derivatives of nitroarginine (**127**) have been described. The nitro group is cleaved by catalytic or electrolytic reduction, but the details of this reaction are not fully understood. Although the cleavage is usually successful when relatively small peptides are involved, by-products can be formed by side reactions of the intermediate N-nitrosoguanidine; with larger peptides this difficulty is sometimes severe. Anhydrous hydrogen fluoride is also reported to cleave the nitroguanidino group and, although the mechanism of this reaction is not understood, it may represent an improvement.

Table 3.1. Protecting groups in common use in peptide synthesis

R	Introduction	Cleavage					
		HBr–AcOH	TFA (cold)	H_2–Pd	Na/NH_3	NaOH	HF
	Carboxyl protection $R^1.CO_2R$						
Me	$NH_2CHRCO_2H/MeOH/SOCl_2$ i.e. $(MeO)_2SO$	S	S	S	complex	C	S
Et	$NH_2CHRCO_2H/EtOH/HCl$	S	S	S	complex	C	S
Bu^t	$NH_2CHRCO_2H/Me_2C{=}CH_2/H_2SO_4/CH_2Cl_2$	C	C	S	?	S	C
CH_2Ph	$NH_2CHRCO_2H/C_6H_6/PhCH_2OH/$ p-$CH_3C_6H_4SO_3H$, water removed azeotropically.	C (slowly)	S	C	C	C	C
p-$CH_2C_6H_4NO_2$	$C_6H_5CH_2OCONHCHRCO_2^-/$ p-$ClCH_2C_6H_4NO_2$, followed by HBr–AcOH.	S	S	C	C	C	C
	Amino protection $R^1N(H)R$						
Bu^tOCO	Bu^tOCON_3, $Bu^tOCO_2C_6H_4$ p-NO_2 or $Bu^tOCOF/NH_2CHRCO_2^-$ (see Exercise 4.6)	C	C	S	S	S	C
$PhCH_2OCO$	$PhCH_2OCOCl/NH_2CHRCO_2^-$ (see Exercise 3.20)	C	S	C	C	S	C
Ph–C_6H_4–CMe_2OCO	via carbonate or azide (compare with Exercise 4.6)	C	C	—	—	S	C
$C_6H_4(CO)_2$ (phthaloyl) (*a*)	$C_6H_4(CO)_2NCO_2Et/NH_2CHRCO_2^-$ (see Exercise 3.23)	S	S	S	complex	complex	S

p-$CH_3C_6H_4SO_2$	p-$CH_3C_6H_4SO_2Cl/NH_2CHRCO_2^-$	S	S	complex	C	S	S
Ph_3C.	$Ph_3CCl/NH_2CHRCO_2^-/Et_2NH$	C	C	C	—	S	C
o-$NO_2C_6H_4S$— (*b*)	o-$NO_2C_6H_4SCl/NH_2CHRCO_2^-$	C	C	—	—	—	—
Hydroxyl protection R^1OR							
Bu^t	$ROH/CH_2{=}CMe_2/H_2SO_4$	C (complex)	C	S	S	S	C
CH_2Ph	(e.g. serine) $BrCH_2CHBrCO_2H/$ $^-OCH_2Ph$/then aminolysis and resolution of N-formyl derivative etc.	C (complex)	S	C	C	S	C
Thiol protection $R^1S.R$							
CH_2Ph	$PhCH_2Cl/RS^-$ in liquid NH_3	S (normal conditions)	S	catalyst poisoning	C	S	C
CH_2CONH_2 (*c*)	$HOCH_2CONH_2/HCl/RSH$	S	S	—	—	S	S
Guanidino protection $R^1NHC(=NR)-NH_2$							
NO_2	$RNH.C(=NH)NH_2/HNO_3-H_2SO_4$	S	S	C	complex	S	C
p-$CH_3C_6H_4SO_2$	p-$CH_3C_6H_4SO_2Cl/RNHC(=NH)NH_2/$ $NaOH$.	S	S	S	C	S	S

Abbreviations: S indicates that the protecting group is stable and C that it is cleaved under the normal conditions of use for the reagents. (*a*) cleaved by the action of NH_2NH_2 etc. (*b*) cleaved by anhydrous HCl. (*c*) cleaved by HgII.

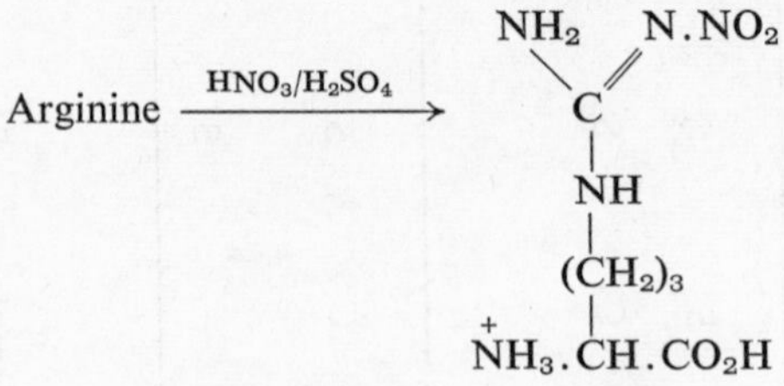

$$\text{Arginine} \xrightarrow{HNO_3/H_2SO_4} \overset{+}{N}H_3.CH(CO_2H).(CH_2)_3.NH.C(NH_2)=N.NO_2$$

(127)

Exercise 3.26

In an attempt to prepare N^{α}-*t*-butoxycarbonyl-N^{δ}-nitroarginine, nitroarginine was refluxed in aqueous sodium bicarbonate in the presence of *t*-butyl-*p*-nitrophenylcarbonate and *t*-butanol. *t*-Butanol was distilled from the reaction mixture and *p*-nitrophenol was removed by ether extraction. Acidification of the remaining aqueous solution gave a substance (A) $C_6H_{10}N_4O_4$ which did not evolve carbon dioxide when treated with concentrated hydrochloric acid. (A) was soluble in aqueous sodium bicarbonate but insoluble in water; it did not give a colour when treated with ninhydrin; its neutralization equivalent was 202; it absorbed in the infrared region at 3268, 1695 and 1600 cm^{-1} (*inter alia*) and, in the ultraviolet, at 270 nm (ϵ = 14,000) and 217 nm (ϵ = 4450). Nitroarginine itself has λ_{max} = 270 nm (ϵ = 15,400); Nitromethylamine, λ_{max} = 230 nm (ϵ = 7000).

When heated in 1M sodium hydroxide, (A) was converted to ornithine. Catalytic hydrogenolysis of (A) gave a crystalline amphoteric substance, (B) $C_6H_{11}N_3O_2$. Account for these observations and deduce possible structures for (A). Suggest further work which might help to determine which of these structures is correct.

Protecting groups: summary

Protecting groups in common use in peptide synthesis are listed in Table 3.1, together with an indication of the means by which they can be introduced into the amino acid moiety and a summary of their stabilities towards various important reagents. It will be apparent, that an appreciable amount of forward planning is necessary for the synthesis of even a simple peptide. Attention to the relative stabilities of the various protecting groups which are to be present at each stage of the synthesis, constitutes an important part of this planning.

CHAPTER FOUR

Chemical Synthesis II

Carboxyl group activation and coupling

The rate-determining step in displacements of the type (**128** → **129**), where X^- is a good leaving group, is the initial attack of the nucleophile on the carbonyl carbon atom. This addition stage is dominated by the electronic and steric effects of the substituents R and X. Electron-withdrawing substituents will tend to increase the polarity $>\overset{\delta+}{C}=\overset{\delta-}{O}$ of the carbonyl group and enhance its reactivity towards nucleophiles; electron-donating substituents will exert the opposite effect. The steric effect of substituents R and X is manifest when these groups inhibit the approach of the nucleophile to the carbonyl group; it is naturally related to the bulk of the nucleophile. These effects account for the well-known diminution in reactivity along the series HCHO > RCHO > RCOR > RCO_2R > $RCONH_2$ > RCO_2^-.

$$R{-}C(=O){-}X \; (+ :B^-) \rightleftharpoons R{-}C(O^-)(B){-}X \rightleftharpoons R{-}C(=O){-}B + X^-$$

(128) **(129)**

Intermediates employed in the preparation of derivatives of carboxylic acids, possess X groups which are electron-attracting and which make good leaving groups. Acyl halides, acyl azides and acid anhydrides are familiar examples. Such derivatives would, of course, be sufficiently reactive for peptide bond formation, but, for an approach to peptide synthesis, other criteria also need to be satisfied. Thus, it must be possible to bring about the activation of the carboxyl group without disturbing the protecting groups already incorporated in the amino acid moiety; the activation and, particularly, the subsequent coupling step must be as near quantitative as possible; side reactions throughout the synthesis must be kept to a minimum, not merely to conserve material but also to avoid the formation of by-products, which may be difficult to remove from the required product; the optical purity of the amino acid residues must be preserved.

Retention of configuration is of sufficient importance to merit a section to itself (p. 96), but it is worth mentioning at this point that even a small degree of racemization at each coupling stage will lead to very impure products. Consider, for example, the synthesis of a pentapeptide, A.B.C.D.E., where the letters refer to α-amino acid residues, by the sequential addition of suitably activated A,B,C and D residues to the terminal residue, E, i.e.

$$\mathrm{D.X + E \longrightarrow D.E + HX}$$

$$\mathrm{C.X + D.E \longrightarrow C.D.E + HX}$$

$$\mathrm{B.X + C.D.E \longrightarrow B.C.D.E + HX}$$

$$\mathrm{A.X + B.C.D.E \longrightarrow A.B.C.D.E + HX}$$

For simplicity, the necessary steps involving protecting groups can be neglected. If L-α-amino acids are used in the synthesis and it is assumed that each of the activated residues, D–A, is incorporated with 80 per cent retention of configuration, the products will possess the following compositions given in Table 4.1.

Already, at the pentapeptide stage, the optical purity is down to 41 per cent. This represents a serious loss of material, particularly since chemical yield has not been taken into account in the calculation. If, say, 75 per cent of the theoretical yield were obtained at each coupling stage, the chemical yields of the peptides would be DE, 75 per cent; CDE, 56·25 per cent; BCDE, 42·19 per cent; ABCDE, 31·64 per cent. The yield of optically pure pentapeptide is thereby reduced to $40{\cdot}1 \times 0{\cdot}316 \approx 12{\cdot}7$ per cent, based on the original weight of E. It is pertinent to recall at this stage that insulin, which is usually considered to be the smallest protein, is composed of two peptide chains, containing twenty-one and thirty residues respectively.

Perhaps even more important than the material loss, is the presence of the unwanted diastereoisomers in the mixture. The above calculation of optical purity assumed that each coupling proceeded symmetrically, i.e. that the chances of a given molecule of the activated amino acid residue reacting with any one of the diastereoisomers present was proportional only to the amount of the diastereoisomer in the mixture. This is undoubtedly an oversimplification, but the picture which emerges is probably

Table 4.1

Peptide	Diastereoisomer composition (per cent)[a]							
DE				LL 80				
CDE		LLL 64				DLL 16		
BCDE	LLLL 51·2		DLLL 12·8		LDLL 12·8		DDLL 3·2	
ABCDE	LLLLL 40·96	DLLLL 10·24	LDLLL 10·24	DDLLL 2·56	LLDLL 10·24	DLDLL 2·56	LDDLL 2·56	DDDLL 0·64
DE				DL 20				
CDE		LDL 16				DDL 4		
BCDE	LLDL 12·8		DLDL 3·2		LDDL 3·2		DDDL 0·8	
ABCDE	LLLDL 10·24	DLLDL 2·56	LDLDL 2·56	DDLDL 0·64	LLDDL 2·56	DLDDL 0·64	LDDDL 0·64	DDDDL 0·16

[a] These percentages are, of course, the individual terms of the binomial expansion $(p + q)^n$. For such an expression, the $(k + 1)^{st}$ term,

$$P_{(n,k)} = \frac{n!}{k!\,(n-k)!}\,p^k q^{n-k}$$

not too distorted. The all L-pentapeptide is therefore contaminated by fifteen diastereoisomers, which are present in amounts varying from 10·24 per cent to 0·16 per cent of the whole. It is extremely difficult to obtain an optically pure entity from such a complex mixture and, since any of the diastereoisomers may possess chemical, physical and biological properties which differ, sometimes dramatically, from those of the optically pure parent compound, the problem is a serious one.

Of the many coupling methods which have been investigated, relatively few have proved satisfactory. Those commonly used in peptide synthesis involve one of four classes of compounds: acyl azides, acid anhydrides, active esters and diimides. Another potentially important method, employing isoxazolium salts, is closely related to the active ester and diimide methods.

Exercise 4.1

N-Tosylglutamic acid reacted with thionyl chloride under reflux to form, in good yield, compound A, $C_{12}H_{12}NSO_4Cl$. Ethyl glycinate was shaken in solution with compound A and a crystalline product, B, $C_{16}H_{20}N_2SO_6$, resulted. B was saponified and then heated with concentrated aqueous ammonia. Sodium–liquid ammonia reduction of the product gave glutaminylglycine. Account for these transformations.

The azide method

Azides (**130**) can be produced readily from α-amino acid esters (**131**) by the action of hydrazine followed by a nitrosating agent.

$$\underset{\textbf{(131)}}{R.CO_2Et} \xrightarrow{NH_2NH_2} R.CO.NH.NH_2 \xrightarrow[-5°C]{HNO_2} \underset{\textbf{(130)}}{R.CO.N_3}$$

---

*Exercise 4.2**

It is interesting to speculate on the mechanism of acyl azide formation by comparing it with the reaction between aliphatic amines and nitrous acid. Remember that the reactive nitrosating agent is dinitrogen trioxide, i.e. the anhydride, O═N—O—N═O.

Azide formation presumably occurs in the following way (**132–133**):

$$O{=}N{-}O{-}N{=}O$$

$$\underset{\textbf{(132)}}{R.CO.NH.NH_2} \rightleftharpoons R.CO.NH.\overset{+}{N}H_2{-}N(O^-){-}O{-}N{=}O \rightleftharpoons$$

$$R.CO.NH.\overset{+}{N}H_2{-}N{=}O \rightleftharpoons R.CO.NH.NH.N{=}O \rightleftharpoons$$

$$R.CO.NH.N{=}N{-}OH \rightleftharpoons \underset{\textbf{(133)}}{[R.CO.N{=}\overset{+}{N}{=}\overset{-}{N} \longleftrightarrow R.CO.\overset{-}{N}{-}\overset{+}{N}{\equiv}N]}$$

The intermediate hydrazides (**132**) are usually quite stable and even crystalline, whereas the acyl azides (**133**) are unstable and are not isolated. They react with the amino component to yield the required peptide (**134**) and hydrazoic acid. The latter is such a weak acid ($pK_a \approx 5$) that it is not necessary to add excess base during the coupling to ensure its neutralization.

$$R.NH.CHR^1.CO.N_3 + NH_2.CHR^2.CO_2Et \longrightarrow \underset{\textbf{(134)}}{R.NH.CHR^1.CO.NH.CHR^2.CO_2Et} + N_3H$$

Occasionally, it is undesirable to expose the peptide to hydrazine, for example, when a phthalyl group is present. In these cases, the peptide hydrazide can be prepared via a suitably protected hydrazide intermediate (e.g. **135**).

$$C_6H_4(CO)_2N.CHR.CO.X + \underset{\textbf{(135)}}{NH_2.CHR^1.CO.NH.NH.CPh_3} \xrightarrow{-HX}$$

$$C_6H_4(CO)_2N.CHR.CO.NH.CHR^1.CO.NH.NH.CPh_3 \xrightarrow{H^+}$$

$$C_6H_4(CO)_2N.CHR.CO.NH.CHR^1.CO.NH.NH_2$$

*Exercise 4.3**†

When N^{α}-benzoylvalyl azide was reacted with valine methyl ester in the cold in an inert solvent, the product was not the desired dipeptide derivative. Saponification of the product gave an acid which possessed a neutralization equivalent of 335 ± 4. Amino acid analysis of this compound yielded a mole equivalent of valine. The presence of a short-lived intermediate [$\nu_{max} \approx 2220\ cm^{-1}$] could be demonstrated spectroscopically during the coupling reaction. Postulate how the unexpected product might have been formed.

† See also appendix.

*Exercise 4.4**†

N^{α}-Benzyloxycarbonylserine azide couples with amino acid esters at −5 to 0°C to give the corresponding dipeptide derivatives. In the absence of added amine, prolonged standing or warming of the azide gives a crystalline compound, $C_{11}H_{12}N_2O_4$. Alternatively, warming or standing the azide in ethyl alcohol gives another derivative, $C_{13}H_{18}N_2O_5$. Suggest probable structures for the unknown compounds and explain how they might have been formed.

† See also appendix.

*Exercise 4.5**†

α,β-Dibenzyloxycarbonylaminopropanoic acid azide gave rise, on standing, to a urea derivative $C_{19}H_{19}N_3O_5$. Postulate a structure for this compound and indicate how it was probably formed.

† See also appendix.

The azide method of coupling is subject to several undesirable side-reactions, one of which is involved in Exercises 4.3*–4.5*. In these examples, the azide undergoes a Curtius rearrangement to produce the corresponding isocyanate (**136**). Of course, isocyanates in general are readily susceptible to nucleophilic addition and this can occur intermolecularly (Exercise 4.3*) or intramolecularly (Exercises 4.4* and 4.5*). Thus, urea derivatives (**137**), which are often similar in properties to the required peptide and therefore difficult to separate from it, may be produced by the attack of the amino component on the isocyanate. The Curtius rearrangement occurs when acyl azides are warmed and hence the importance of keeping the reaction mixture cold.

$$R{-}\overset{O}{\overset{\|}{C}}{-}\ddot{N}{=}\overset{+}{N}{=}\overset{-}{N} \longleftrightarrow R{-}\overset{O}{\overset{\|}{C}}{-}\overset{-}{\ddot{N}}{-}\overset{+}{N}{\equiv}N \longrightarrow \left[R{-}\overset{O^-}{\overset{|}{C}}{=}N{:}\right] + N_2 \longrightarrow$$

$$\underset{(\mathbf{136})}{R{-}N{=}C{=}O} \xrightarrow[NH_2.R^1]{} \underset{(\mathbf{137})}{R.NH.\overset{O}{\overset{\|}{C}}.NH.R^1}$$

Another impurity, which is sometimes formed during azide coupling reactions, is the primary amide (**138**) analogous to the acyl hydrazide. It is thought that the amide does not arise due to a rearrangement of the azide itself, because, if the azide is prepared from the acid chloride and sodium azide, amide formation is not detected. More likely, the amide is formed directly from the hydrazide. Amide formation is at a minimum when the water content of the reaction mixture is low and, consequently, *t*-butyl nitrite and nitrosyl chloride, which can be used in anhydrous solvents, have been advocated as preferred nitrosating reagents.

$$R{-}\overset{O}{\overset{\|}{C}}{-}NH.NH_2 \dashrightarrow R{-}\overset{O}{\overset{\|}{C}}{-}NH{-}N{=}N{-}OH \xrightarrow{-N_2O} \underset{(\mathbf{138})}{R.CO.NH_2}$$

The azide coupling remains important in spite of its several shortcomings because it seems to be entirely free from racemization.

Exercise 4.6

t-Butylchloroformate is relatively unstable and cannot be stored. It is therefore hardly suitable for the preparation of *t*-butoxycarbonylamino acids. The corresponding fluoride is so stable that it can be distilled, but, under conditions of controlled pH, it reacts with amino acids to give excellent yields of the *t*-butoxycarbonyl derivatives. Account for the difference in reactivity between the two alkoxyacyl halides.

The *t*-butoxycarbonyl protecting group is also introduced conveniently by the action of *t*-butylazidoformate. Suggest a method for the preparation of this reagent.

The anhydride method

The activation of carboxylic acids by the formation of anhydrides has commonplace familiarity and it is not surprising that the activation of α-amino acid derivatives has been undertaken in this way. Symmetrical anhydrides (**139**) have the disadvantage that only half of the constituent carboxylic acid can be utilized.

$$\underset{\textbf{(139)}}{(R.CO)_2O} + NH_2.R^1 \longrightarrow R{-}C(O^-)({}^+NH_2.R^1){-}O{-}C(=O){-}R \dashrightarrow R{-}CO.NH.R^1 + R.CO_2H$$

This difficulty can be overcome by the use of unsymmetrical (i.e. mixed) anhydrides of the type $R.CO.O.CO.R^1$. For example, the mixed anhydride between formic and acetic acids (**140**) has been used as a formylating agent. The relative reactivities of the respective carbonyl groups is indicated by the fact that, even with acetic acid as solvent, no acetylation occurs.

$$\underset{\textbf{(140)}}{H.CO.O.CO.CH_3} + NH_2R \rightleftharpoons H.CO.NH.R + CH_3.CO_2H$$

Mixed anhydrides (**141**) can be prepared easily by the reaction between a tertiary base salt of the carboxyl component (**142**) and the appropriate acid chloride (**143**). These anhydrides tend to disproportionate to a mixture of the symmetrical anhydrides and so they are prepared at low temperature and used immediately. They are not isolated.

$$R.CO.NH.CHR^1.CO_2^-\overset{+}{H}NEt_3 + R^2.CO.Cl \longrightarrow$$
(142) **(143)**

$$R.CO.NH.CHR^1.CO.O.CO.R^2 + \overset{+}{N}Et_3HCl^-$$
(141)

$$\textbf{(141)} + NH_2.CHR^3.CO_2Et \longrightarrow R.CO.NH.CHR^1.CO.NH.CHR^3.CO_2Et + R^2.CO_2H$$

---

*Exercise 4.7**†

N-Protected α-methylalanine fails to react with α-methylalanine methyl ester under a variety of standard coupling conditions. *N*-Benzyloxy-carbonyl-α-methylalanyl-α-methylalanine can be made, however, via the mixed anhydride formed with pivalic acid (trimethylacetic acid). When the dipeptide was reacted with pivaloyl chloride in the presence of triethylamine, a crystalline compound, $C_{16}H_{20}N_2O_4$, was obtained in quantitative yield. This compound reacted readily with methyl α-methylalanine ester at 100°C to yield the *N*-benzyloxycarbonyl tripeptide ester quantitatively. Trace the reactions involved.

† See appendix.

---

*Exercise 4.8**†

The mixed anhydride formed from *N*-tosylglycine and *sec*-butyl-chloroformate in the presence of an equimolar amount of pyridine, reacted with aniline to give at least ten products, among which could be identified: tosylglycine, its aniline salt and anilide: *N*-tosyl(*N*-tosyl-glycyl)glycine; 1,4-ditosyl-2,5-dioxopiperazine and *N*-*sec*-butoxycarbonyl-*N*-tosylglycine. Account for this behaviour and postulate possible structures for the other products.

† See appendix.

It is interesting to consider the model reactions between *N*-methylaniline and mixed carbonic–carboxylic anhydrides involving hydrocinnamic (**144**; R = $Ph.CH_2CH_2$), cinnamic (**144**; R = Ph.CH=CH) and sorbic (**144**; R = CH_3.CH=CH.CH=CH) acids. Clearly, two products can be formed, the amide (**145**) and the urethane (**146**), dependent on the point of attack of the amine on the anhydride.

R—C(=O)—O—C(=O)—OEt (**144**) + HNMePh → R.CO.NMePh + EtOH + CO_2 (**145**)

R—C(=O)—O—C(=O)—OEt (**144**) + HNMePh → $R.CO_2H$ + EtO.CO.NMePh (**146**)

The relative proportions of the products are informative:

Anhydride	Products (per cent)			
	Amide	Urethane	Acid	Unreacted amine
Hydrocinnamic	75	8·4	15	5
Cinnamic	63	17	20	5
Sorbic	49	20	29	4

The reactivity of the carbonyl group derived from these acids diminishes with increasing conjugation. Similarly, the saponification of esters of α,β-unsaturated carboxylic acids is slower than the saponification of their saturated analogues. The duality of the aminolysis reaction in the above examples, focusses attention on one of the difficulties in the use of mixed anhydrides in peptide synthesis, but at the same time, it suggests a solution to the problem.

Since mixed anhydrides possess two reactive carbonyl groups there is a danger in peptide synthesis that nucleophilic attack will occur at the 'wrong' centre. In this case, the product will be the acyl derivative of the amino component. It might or might not be difficult to separate this amide from the required material, but the formation of the amide represents a material loss. This difficulty can be avoided if the second carboxylic acid component of the mixed anhydride is selected so that the carbonyl group which it contributes is relatively unreactive. Both steric and electronic effects may be important in achieving this objective. The desired peptide-forming reaction usually takes place preferentially when the second carboxyl component has an α or β-branched chain as, for example, in isobutyric acid mixed anhydrides, although, in particularly unfavourable cases, the undesirable reaction can predominate. This is so with α-methylalanine couplings (Exercise 4.7*) in which steric hindrance is severe, but in this instance, the difficulty can be overcome by using a mixed pivalic acid anhydride (**147**). The carbonyl group derived from the pivalic acid moiety is electronically deactivated by the (+*I*) effect of the *t*-butyl group and it is severely hindered; even in the case of α-methylalanine, attack occurs entirely at the other carbonyl group.

$$R.CO.NH.CHR^1.CO.O.CO.CMe_3$$

(147)

There are two further disadvantages to the use of mixed anhydrides of carboxylic acids in peptide synthesis. As the coupling proceeds, the carboxylic acid is released from the mixed anhydride and, unless extra base is present to neutralize it, a salt is formed with the amino component. Consequently, the coupling must be carried out in the presence of excess tertiary base which, on other accounts, is undesirable (p. 99). In addition, the carboxylic acid produced in the coupling can be difficult to separate from the peptide product.

A means of avoiding both these difficulties is also suggested by the model reactions considered above. When mixed carboxylic–carbonic anhydrides react with amines, only in the urethane-forming reaction does a potentially troublesome impurity remain in the reaction mixture. The amide-forming reaction gives, as a by-product, the carbonic ester which decomposes rapidly to carbon dioxide and ethanol. Mixed carboxylic–carbonic anhydrides are therefore very convenient to use in peptide synthesis. They can be readily prepared by reacting an alkyl chloroformate, instead of an acyl chloride, with a tertiary base salt of the carboxyl component. Ethyl chloroformate and isobutyl chloroformate are commonly used in this way.

---

*Exercise 4.9**

Reaction half-times for the aminolysis of substituted phenyl esters of *N*-benzyloxycarbonyl-L-phenylalanine (Z.Phe.OX), by excess benzylamine in aqueous dioxan solution at 25°C, are recorded below, together with the Hammett σ functions for the corresponding substituents. Consider whether the relative rates of aminolysis can be explained satisfactorily in terms of the electronic effects of the various substituents, and calculate an approximate half-life for the aminolysis of the biphenyl ester. What easily measured physical property could be employed to indicate the probable relative rate of aminolysis of any given ester?

X	$-C_6H_4-C_6H_5$ (p)	$-C_6H_4Cl$ (p)	$-C_6H_4F$ (m)
$t^{1/2}$(min)	1000	700	390
σ	$-0{\cdot}01$	$+0{\cdot}227$	$+0{\cdot}337$

X	$-C_6H_4Cl$ (m)	$-C_6H_4COCH_3$ (p)	$-C_6H_4NO_2$ (p)
$t^{1/2}$(min)	360	157	5·3
σ	$+0{\cdot}373$	$+0{\cdot}502$	$+0{\cdot}778$

The active ester method

Since peptide bond formation, even from amino acid methyl esters, is thermodynamically favoured, it might be expected that this type of reaction would be even more feasible if the methyl moiety were replaced by stronger electron-withdrawing groups. Thus, cyanomethyl esters (**148**), readily formed from a tertiary base salt of the carboxyl component and α-chloroacetonitrile in an inert solvent, are aminolysed rapidly and have been employed in peptide synthesis. Esters like this, which have enhanced rates of aminolysis, are referred to, somewhat loosely, as active esters.

$$R.CO_2^-H\overset{+}{N}Et_3 + Cl.CH_2.CN \longrightarrow \underset{\mathbf{(148)}}{R.CO_2.CH_2.CN} + H\overset{+}{N}Et_3Cl^-$$

$$\mathbf{(148)} + NH_2.R^1 \longrightarrow R.CO.NH.R^1 + HO.CH_2.CN$$

Phenyl esters provide much more scope for the introduction of substituents to modify their properties and substituted phenyl esters are very commonly employed in peptide synthesis. Some examples are considered in Exercise 4.9*. Since benzylamine is present in excess, the reactions in this exercise will obey first order kinetics, and the familiar equation:

$$\frac{d(a-x)}{dt} = k(a-x)$$

where a is the initial concentration of the ester and x its concentration at time t, will apply. Integration gives:

$$\int_0^t dt = \frac{1}{k}\int_0^x \frac{dx}{a - x}$$

$$\therefore k = \frac{2 \cdot 303}{t} \log\left(\frac{a}{a - x}\right)$$

It is easily seen from this equation that:

$$\log\left(\frac{k}{k_H}\right) = \log\left(\frac{t_H^{1/2}}{t^{1/2}}\right)$$

where k and k_H are used in the sense of the Hammett equation (p. 215) and $t_H^{1/2}$ and $t^{1/2}$ are the half reaction times of the unsubstituted and substituted phenyl esters respectively.

If the difference between the relative rates of aminolysis can be attributed entirely to the electronic effect of the substituents, then:

$$\log\left(\frac{t_H^{1/2}}{t^{1/2}}\right) = \rho\sigma$$

$$\text{i.e. } -\log t^{1/2} = \rho\sigma - \log t_H^{1/2}$$

This equation can be checked by plotting $\log t^{1/2}$ against σ. The plot gives a reasonable approximation to a straight line which indicates the dominance of the electronic effect. A value for ρ of approximately $+3$ is obtained from the slope of the line and, since this is quite a high value for the reaction constant, the reaction seems to be fairly sensitive to electronic changes of this type. The half reaction times for the unsubstituted phenyl ester ($\sigma = 0$) and for the biphenyl ester ($\sigma = -0 \cdot 01$) can also be obtained from the plot. [Note: Remember that ρ is characteristically positive in reactions which are favoured by withdrawal of electrons from the active site and that, for the dissociation of the ring-substituted anilinium ion in water at 25°C, $\rho \approx +2 \cdot 7$ (p. 63)].

Although, in the examples considered above, the Hammett substitution constant provides a reasonably accurate measure of the relative rates of aminolysis of the esters, in other cases the correlation is not so good. In particular, entropic factors can be important, especially when *ortho* substituents are present and, in these instances, the Hammett equation is often not applicable. The issue can be further complicated by catalysis. For example, the rate of aminolysis of cyanomethyl esters is appreciably increased in the presence of acetic acid. Bifunctional catalysts, which possess both a weakly basic and a weakly acidic group, for example 1,2,4-triazole, also accelerate the aminolysis of active esters and their use in peptide synthesis has been advocated (compare with enzymic catalysis, p. 174 *et seq*).

Disregarding these anomalies, an approximate indication of the likely reactivity of a given ester can usually be obtained from the pK_a value of the phenol from which it is derived. Substituents which enhance the reactivity of the carbonyl towards nucleophiles, will also tend to increase the acidity of the phenol, i.e. they will increase the stability of the substituted anion relative to PhO^-. This is borne out fairly well in the examples of Exercise 4.9*, in which the reported pK_a values of the phenols are *p*-chloro, 11·60; *m*-fluoro, 11·88; *m*-chloro, 11·50; *p*-acetyl, 10·62; *p*-nitro, 9·27. A plot of pK_a against σ for these esters approximates to a straight line, the slope being almost the same as the value of ρ derived from the aminolysis reaction (Figure 4.1).

It does not follow that the most acidic phenols will give the best active esters. Side reactions might be dominant if the ester is too reactive. For example, the ester formed from *N*-benzyloxycarbonyl-L-phenylalanine and picric acid has not been isolated because it disproportionates spontaneously to form *N*-benzyloxycarbonyl-L-phenylalanine anhydride.

Phenols which possess pK_a values in the range 9–10 seem to be particularly suitable for the preparation of active esters for peptide synthesis. The two best known types of active ester are derived from *p*-nitrophenol and from 2,4,5-trichlorophenol (pK_a 9·45, $t^{1/2}$ = 3·3, where $t^{1/2}$ has the significance given to it in Exercise 4.9*). Despite the apparent difference in the pK_a values, the trichlorophenyl esters seem to couple consistently faster than the *p*-nitrophenyl esters, but this is not a significant advantage. On the other hand, some difficulty is occasionally encountered with *p*-nitrophenyl esters because of the persistence of the *p*-nitrophenol released in the coupling reaction. If subsequent stages involve hydrogenolysis, the *p*-nitrophenol is converted to coloured impurities. Trichlorophenol can usually be separated clearly from peptide material. It is fair to comment, however, that a great number of successful syntheses have been accomplished using *p*-nitrophenyl esters, without difficulties of this type.

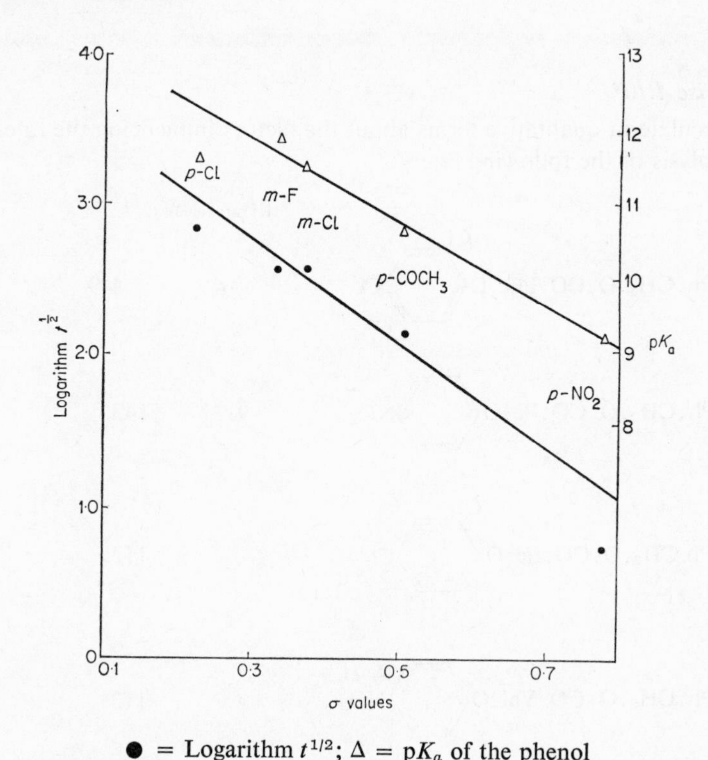

● = Logarithm $t^{1/2}$; Δ = pK_a of the phenol

Figure 4.1. The aminolysis of substituted phenyl esters of *N*-benzyloxycarbonyl-L-phenylalanine by excess benzylamine in aqueous dioxan at 25°C (see Exercise 4.9*). Data taken from the work of J. Pless and R. A. Boissonas *Helvetica Chimica Acta*, 1963, **46**, 1609

The significance of electronic effects has been stressed throughout this section but steric considerations can also be important. Some cases which illustrate this are dealt with in Exercise 4.10*. The relatively slow rates of aminolysis in examples 2–5 are accounted for by the now familiar steric effects created by the β-branched side-chains of isoleucine and valine; in examples 6 and 7, the hindrance originates in the *ortho* substituted ester moiety. It is apparent in examples 6 and 7 that the hindrance increases in severity with the size of the *ortho* substituent, i.e. I > Br. As shown by the trichlorophenyl ester, example 1, the presence of *ortho* substituents is not always so serious.

*Exercise 4.10**

Speculate in qualitative terms about the factors influencing the rates of aminolysis of the following esters:

	$pK_{a(phenol)}$	$t^{1/2}$
1.‡ Ph.CH_2.O.CO.Phe.O-(2,4,6-trichlorophenyl) [Cl, Cl, Cl]	9·45	4·9
2.‡ Ph.CH_2.O.CO.Ile.O-C_6H_4-NO_2	9·27	143
3.‡ Ph.CH_2.O.CO.Ile.O-(2,4,6-trichlorophenyl) [Cl, Cl, Cl]		112
4.‡ Ph.CH_2.O.CO.Val.O-C_6H_4-NO_2		113
5.‡ Ph.CH_2.O.CO.Val.O-(2,4,6-trichlorophenyl) [Cl, Cl, Cl]		95·5
6.† Ph.CH_2.O.CO.Phe.O-(2,4,6-tribromophenyl) [Br, Br, Br]	8·43	72
7.† Ph.CH_2.O.CO.Phe.O-(2,4,6-triiodophenyl) [I, I, I]	8·39	353

† Reaction as in Exercise 4.9*.
‡ Solvent: dioxan; otherwise conditions as in Exercise 4.9*.

*Exercise 4.11**

Although catechol is not much more acidic than phenol, the reactivity of catechol esters of *N*-protected amino acids is similar to that of the corresponding *p*-nitrophenyl esters. Speculate on the reasons for this high order of reactivity.

The presence of certain *ortho* substituents in phenyl esters of amino acids undoubtedly facilitates aminolysis. Esters which exhibit this effect include those derived from catechol, and from oxine (8-hydroxyquinoline). The acidity of these phenols is rather low and their reactivity cannot be accounted for in these terms. It is interesting to compare the reactivity of *N*-benzyloxycarbonylvaline esters derived from oxine, *p*-nitrophenol and α-naphthol. The first two examples couple rapidly with glycine ethyl ester in dioxan at 20°, whereas the α-naphthol ester hardly reacts at all. Since the coupling obeys the second-order rate equation, the enhanced reactivity of the oxine ester is not due to a catalytic effect of the liberated phenol. Another important example of this type of reactivity is provided by esters of *N*-hydroxypiperidine.

It is postulated that the reactivity of these esters arises from the presence in the ester moiety of a group capable of accepting a proton from the attacking amino group. Structures **149–151** indicate the type of transition state which is envisaged. Esters of *N*-hydroxysuccinimide, which are very useful in peptide synthesis because of the solubility of the liberated *N*-hydroxysuccinimide in water, possibly owe their reactivity, at least in part, to a similar effect (**152**). The importance of these various esters will be further stressed in the section which deals with racemization.

The preparation of active esters derived from acidic phenols is best accomplished by the use of a diimide reagent. This type of reaction will be discussed in the next section. Esters from the less acidic phenols can be prepared by the action of the phenol on a mixed anhydride of the *N*-protected amino acid, although other approaches may be preferred. One of the attractions of the active ester approach is that the esters can be prepared and isolated in an optically pure, generally crystalline form quite independently of the coupling reaction.

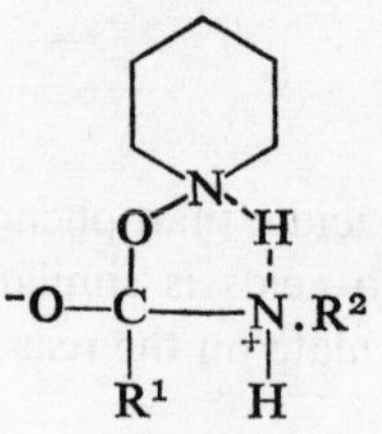

(149) 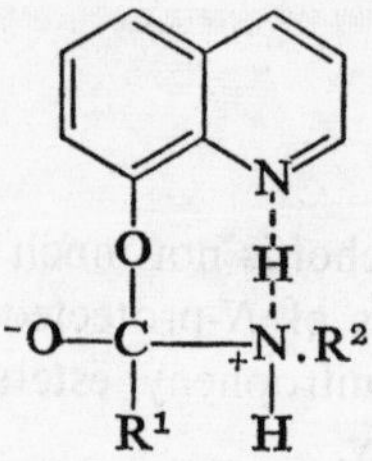 (150)

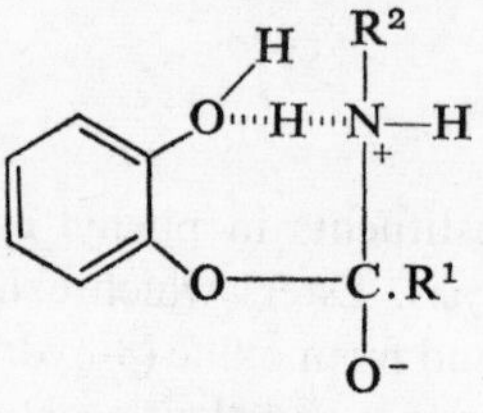

(151) 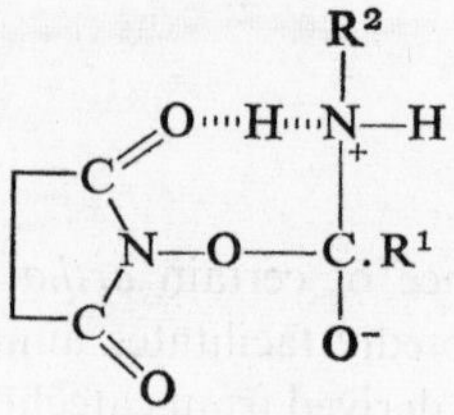 (152)

Exercise 4.12

Esters of the form Ph.CH_2.O.CO.NH.CHR.CO.O.NH.CO.CMe_3 are aminolysed readily by amino acid esters to give peptides. However, the ester from *N*-benzyloxycarbonylproline does not react in this way, but gives *N*-benzyloxycarbonylproline and a urea derivative of the amino acid ester. Account for its behaviour.

The diimide method

Probably the most widely used reagent in peptide synthesis is *N,N'*-dicyclohexylcarbodiimide (153). The addition of this reagent to a mixture of amino-protected and carboxyl-protected α-amino acids, dissolved in an inert solvent, results in the formation of the corresponding protected peptide and *N,N'*-dicyclohexylurea. Esters of moderately acidic to strongly acidic phenols, i.e. active esters, can be made in a similar way from the phenol and the *N*-protected amino acid.

$$R.NH.CHR^1.CO_2H + NH_2.CHR^2.CO_2R^3 + C_6H_{11}{-}N{=}C{=}N{-}C_6H_{11} \longrightarrow$$

(153)

$$R.NH.CHR^1.CO.NH.CHR^2.CO_2R^3 + C_6H_{11}NH.CO.NHC_6H_{11}$$

N,N′-Dicyclohexylcarbodiimide is a weak base which forms a salt, for example, with dry hydrogen chloride. So far as reactivity is concerned, it resembles other cumulative double bond systems involving heteroatoms (e.g. **154–156**), in that it is susceptible to nucleophilic addition. With *N,N′*-dicyclohexylcarbodiimide, the addition of amines, alcohols and most phenols requires forcing conditions, whereas the peptide coupling reaction occurs readily in the cold. This distinction has the practical consequence that side-chain hydroxyl functions do not need to be protected during diimide couplings.

$$R{-}N{=}C{=}S \;\textbf{(154)} \xrightarrow{NH_3} R.NH.CS.NH_2$$

$$R{-}N{=}C{=}O \;\textbf{(155)} \xrightarrow{NH_3} R.NH.CO.NH_2$$

$$R^1R^2C{=}C{=}O \;\textbf{(156)} \xrightarrow{NH_3} R^1R^2.CH.CO.NH_2$$

It is reasonably well established that the diimide reagent reacts initially with the carboxyl component to form the aminoacyl isourea (**157**). The reaction occurs readily in non-polar solvents and it is therefore unlikely to involve free ions, but probably proceeds via an ion-pair transition state (**158**). Thus, carboxylic acids react in order of increasing acidity.

$$\left[R^1.NH.CH.R^2.C(O^{(1/2)-})(O^{(1/2)-}) \cdots HO.C(=O).CH.R^2.NH.R^1\right] \quad C_6H_{11}{-}\overset{+}{N}H{=}C{=}N{-}C_6H_{11}$$

(158)

$$\downarrow$$

$$R^1.NH.CHR^2.CO.O{-}C(=N{-}C_6H_{11}){-}NH{-}C_6H_{11}$$

(157)

The acyl isourea intermediate is effectively an active ester which undergoes aminolysis readily in the normal way. Again it seems possible that a group (the imido nitrogen) may be fortunately placed to assist in the removal of the proton from the attacking amine. The formation of active esters from acidic phenols is similar.

---

*Exercise 4.13**

A major by-product of the *N,N'*-dicyclohexylcarbodiimide coupling reaction between *N*-benzyloxycarbonylvaline and various peptide derivatives has the empirical formula $C_{26}H_{38}N_3O_4$. When this component is hydrolysed, valine and *N,N'*-dicyclohexylurea can be identified amongst the products. Propose a structure for the unknown compound and suggest how it might have been formed.

In some sterically hindered cases a rearrangement of the aminoacyl isourea competes strongly with the aminolysis reaction. The acyl urea (**159**) formed by the rearrangement is not readily susceptible to aminolysis and it remains as an impurity. When an asymmetric diimide is used, the acyl moiety becomes attached to the least basic nitrogen atom. In the absence of amino component, the symmetrical anhydride of the carboxylic acid is also prominent in the products and this reaction has formed the basis for a preparative route to anhydrides.

R.NH.CHR¹.CO—O
N=C—NH →
R.NH.CHR¹.CO
N.CO.NH
(**159**)

Other side reactions are sometimes observed with the diimide reagent. The most important of these concerns the side-chain amide groups of asparagine and glutamine which are susceptible to dehydration to the corresponding nitriles (**160**, $n = 1$ or 2) during attempts to activate their carboxyl groups prior to coupling. Due care must be taken in the incorporation of these residues into peptides.

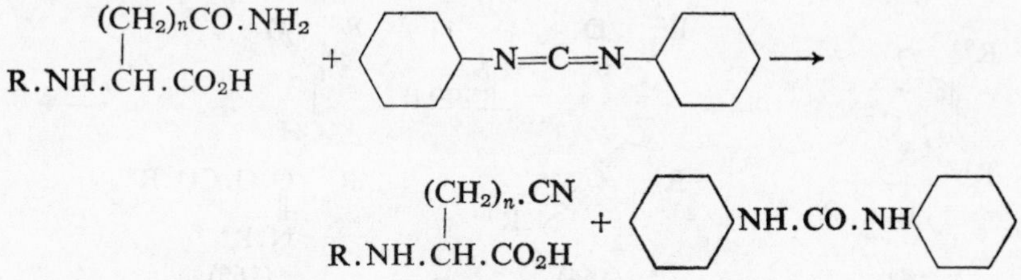

$$R.NH.CH(CO_2H).(CH_2)_nCO.NH_2 + C_6H_{11}{-}N{=}C{=}N{-}C_6H_{11} \longrightarrow$$

$$R.NH.CH(CO_2H).(CH_2)_n.CN + C_6H_{11}NH.CO.NHC_6H_{11}$$

(160)

One drawback to the diimide method is that the urea by-product tends to resemble the peptide solubility-wise and this can make the isolation of the peptide difficult. In critical cases, the difficulty can be overcome by the use of a diimide which carries a polar group to render the derivatives water-soluble (e.g. **161**).

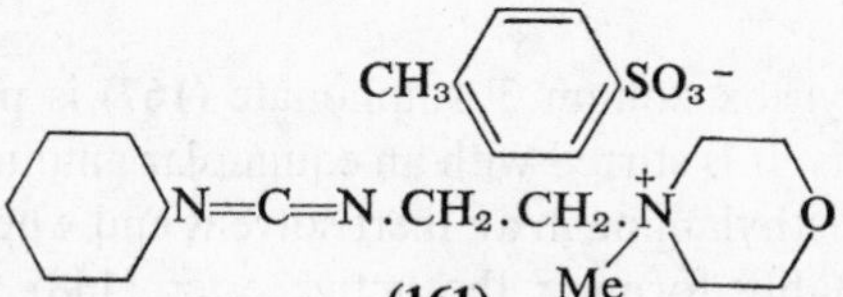

$$CH_3C_6H_4SO_3^-$$

$$C_6H_{11}N{=}C{=}N.CH_2.CH_2.\overset{+}{N}(Me)C_4H_8O$$

(161)

Other compounds which contain cumulative double bonds, for example, the ketenimine (**162**), have also been used in peptide synthesis. For various reasons, these other coupling reagents have not provided serious competition to *N*,*N'*-dicyclohexylcarbodiimide, but one particular technique, the isoxazolium method, merits separate attention.

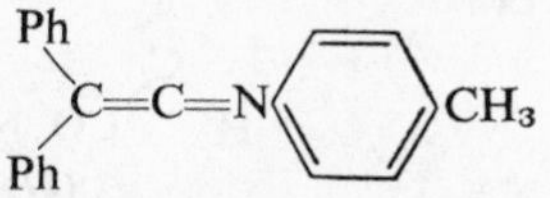

$$Ph_2C{=}C{=}N{-}C_6H_4{-}CH_3$$

(162)

The isoxazolium method

Isoxazolium salts which bear a hydrogen atom at C^3 (**163**), rearrange smoothly under basic conditions to give α-ketoketenimes (**164**). These compounds react with carboxylic acids like other cumulative double bond structures except that the initial product (**165**) tautomerizes and rearranges rapidly to the active ester (**166**):

(163) (164) (165)

(166)

N-Ethyl-5-phenylisoxazolium 3′-sulphonate (**167**) is particularly useful in peptide synthesis. It is stirred with an equimolar amount of the carboxyl component and triethylamine in an inert solvent and when the reagent has dissolved, presumably forming the active ester (**168**), the amino component is added. Coupling is rapid and the by-product (**169**) is water-soluble and easily separated from the product.

(167) (168)

(169)

*Exercise 4.14**

N-Methyl-5-phenylisoxazolium salts react in the above manner with acetate in aqueous solution to form the acetate active ester similar to (**168**) [Infrared: 3470 cm^{-1}, —CONH—; 1770 cm^{-1}, $-\overset{|}{C}=\overset{|}{C}-O-\underline{CO}-$; 1678 cm^{-1}, —CONH—; 1634 cm^{-1}, $-\underline{\overset{|}{C}=\overset{|}{C}}.CONH-$; 1515 cm^{-1}, —CONH—; 1190 cm^{-1}, $-\overset{|}{C}-\overset{|}{C}-\underline{O-CO}-$; ultraviolet: $\lambda_{max} = 267$ nm ($\epsilon = 18{,}700$) CH_2Cl_2, similar to cinnamamide]. In ethanolic solution, the active ester rearranges spontaneously to a new compound, $C_{12}H_{13}NO_3$ [Infrared: no bands in NH, OH range; broad band 1700 cm^{-1}; ultraviolet: λ_{max} 242 nm ($\epsilon = 12{,}300$); 314 nm ($\epsilon = 5100$), similar to acetophenone]. Consider, by analogy with the diimide reaction, what this by-product might be and postulate how the original reagent might be modified to avoid its formation.

The rearrangement considered in Exercise 4.14* involves the migration of the acyl group in the active ester to the imino nitrogen (**170** → **171**). Written in this form, it is immediately seen that the rearrangement is similar to the formation of the substituted acyl urea by-product in the diimide coupling reaction. The rearrangement is much less troublesome when the *N*-ethylisoxazolium derivative is used instead of the *N*-methyl compound, whilst the enol ester (**170**, $R^2 = Bu^t$), formed from the *N-t*-butylisoxazolium derivative, is quite stable from this point of view.

$$R^1.CH_2.\overset{O.\overset{O}{\overset{\|}{C}}.R^3}{\overset{|}{C}}=N.R^2 \longrightarrow R^1.CH_2.\overset{O}{\overset{\|}{C}}-\overset{CO.R^3}{\overset{|}{N}}.R^2$$

(170) **(171)**

2-Ethyl-7-hydroxybenzisoxazolium (**172**), is especially noteworthy in that its salts react with *N*-protected amino acid sodium salts at pH 4·5 to give the substituted catechol ester (**173**). The potential importance of catechol esters was indicated in an earlier section; it will become even more evident in the next.

OH O. $\overset{+}{N}$.Et $\xrightarrow{R^1.CO_2H}$ OH O.CO.R^1 CO.NHEt ----→ O.CO.R^1 OH CO.NHEt

(172) (173)

Racemization

*Exercise 4.15**

The principal mechanism by which racemization occurs during peptide synthesis has become apparent as a result of numerous related observations of which the following are a few:

(a) *N*-Acetyl-α-amino acids are rapidly racemized in the presence of even catalytic amounts of acetic anhydride. *N*-Acetylproline, however, does not racemize under these conditions.

(b) When *N*-benzoylglycyl-L-phenylalanine *p*-nitrophenyl ester in solution in dichloromethane is treated with an equimolar amount of triethylamine, the optical rotation of the solution is reduced by a half after 50 minutes at room temperature. *N*-Phthalyl-L-phenylalanine *p*-nitrophenyl ester is much more stable under these conditions.

(c) If the coupling reaction between *N*-benzoyl-L-leucine *p*-nitrophenyl ester [ν_{max} = 1776 cm^{-1}] and glycine ethyl ester hydrochloride in the presence of *N*-methylpiperidine is interrupted before it can go to completion, the protected dipeptide and compound A can be isolated. Compound A [m.p. = 54·5–55°, $C_{13}H_{15}NO_2$, ν_{max} = 1832 and 1664 cm^{-1} in $CHCl_3$] will react with ethyl glycinate to give the protected dipeptide. Virtually all of the racemization observed in the original coupling can be accounted for in terms of the formed dipeptide and Compound A. Some recovered *p*-nitrophenyl ester was almost racemate-free.

Deduce a mechanism by which racemization could occur which would account for these observations.

At first, it is tempting to speculate that the occurrence of racemization during peptide synthesis might be a consequence of the direct ionization of the α-hydrogen atom, but the type of evidence presented in Exercise 4.15* rapidly dispels the idea that racemization is this simple. Thus the α-hydrogen atom in the phthalyl derivative (Exercise 4.15(b)*) should be more acidic than in the dipeptide derivative, but, whereas the latter racemizes rapidly, the phthalyl compound is relatively resistant to racemization. It is also clear (Exercise 4.15(a)*) that the presence of a H-atom on the α-nitrogen is necessary for racemization to occur. The evidence of Exercise 4.15(c)* leads intuitively to the hypothesis that the necessity for the presence of this hydrogen atom is related in some way to the formation of compound A and that this compound should provide the key to the problem.

Compound A is, in fact, 4-isobutyl-2-phenyloxazolone (**174**; R = Ph) and it is relatively easy to visualize how it is formed from the *p*-nitrophenyl ester (**175**; R = Ph). Under basic conditions, the oxazolone is in equilibrium with the enolate form (**176**), which is an aromatic heterocyclic; hence the racemization. Oxazolone formation can also be expected in carboxyl-activated peptides, as in Exercise 4.15(b)* (**174**; R = Ph.CO.NH.CH_2), and in carboxyl-activated acetylamino acids, as in Exercise 4.15(a)* (**174**; R = CH_3). In the latter example, the carboxyl group was probably activated by the formation of a mixed anhydride.

H–B; N, $CH_2.CHMe_2$; R.C, CH; O, C=O; OC_6H_4p-NO_2
(175)

⇌

N, $CH_2.CHMe_2$; R.C, CH; O—CO
(174)

⇌

N, $CH_2.CHMe_2$; R.C, C; O—C; O^-
(176)

Studies with isolated oxazolones confirm that they are themselves susceptible to aminolysis to form peptides, and that they are also attacked by the *p*-nitrophenate ion to form *p*-nitrophenyl esters. This is hardly surprising in view of the similarity between oxazolones and some of the active esters considered in earlier sections of this chapter. Under the conditions of peptide synthesis, racemization of the oxazolone through the enolic form, is normally much more rapid than the ring-opening reactions. Oxazolones are well characterized spectroscopically, but their presence in the racemizing mixture cannot always be demonstrated; presumably, in these cases, the equilibrium concentration of oxazolone is low.

Racemization associated with peptide-bond formation is, therefore, a complex phenomenon in which the following elements can be distinguished.

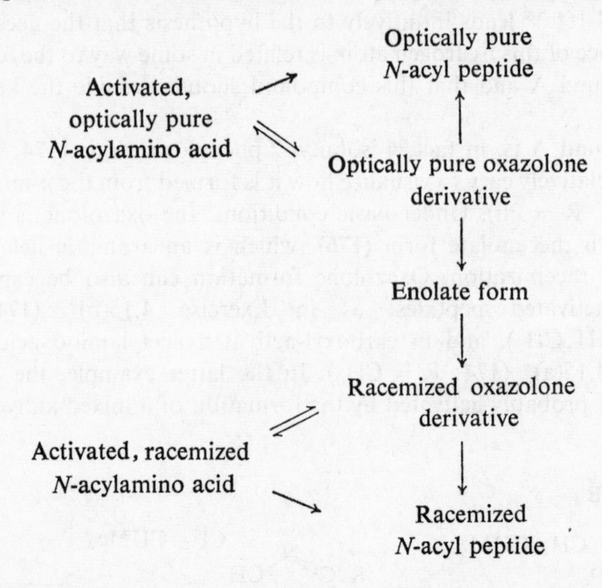

According to this mechanism, only carboxyl-activated amino acids and peptides can racemize readily, and this is confirmed by experience. One other important finding is that activated amino acids which bear urethane-type protecting groups do not form oxazolones and hence do not racemize in this way. Both of these points must be borne in mind when the synthesis of an optically active peptide is contemplated.

---

*Exercise 4.16**

Account for the following observations: *N*-Benzyloxycarbonyl-*S*-benzyl-L-cysteine *p*-nitrophenyl ester in solution in *N*,*N*-dimethylformamide slowly racemizes at room temperature. The racemization is accelerated by the addition of triethylamine. A clue to the mechanism of this reaction is provided by the formation of *N*-benzyloxycarbonyl-*S*-benzyl-DL-cysteine thiobenzyl ester when *N*-benzyloxycarbonyl-*S*-benzyl-L-cysteine is heated at 100°C in the presence of triethylamine.

---

*Exercise 4.17**

Explain the following: *N*-Benzyloxycarbonyl-L-phenylglycine *p*-nitrophenyl ester racemizes readily in the presence of base and the corresponding *N*-phthalyl derivative racemizes even faster. For example, a solution of the phthalyl derivative in tetrahydrofuran containing 1 per cent triethylamine, shows a 50 per cent drop in optical rotation in less than 1 minute.

Two other mechanisms by which racemization can occur are of less general importance but have been observed in certain instances. The first mechanism, which implicates a β-elimination step (**177** $\rightarrow$ **178**), has been encountered before in a different guise (Exercise 3.12); it is encountered once more in Exercise 4.16*. The second mechanism is direct ionization. It is particularly favoured in the case of phenylglycine (Exercise 4.17*) because the anion produced by the loss of a proton is stabilized by delocalization involving the benzene ring; as expected, ionization is even more favoured in the case of the phthalyl derivative (**179**).

It is interesting to see that racemization by each of the three known mechanisms can be measured by the use of suitable model compounds. For example, using the sterically hindered base *N*,*N*-diisopropylethylamine and the *p*-nitrophenyl esters of *N*-benzoyl-L-leucine (oxazolone mechanism), *N*-benzyloxycarbonyl-*S*-benzyl-L-cysteine (β-elimination mechanism) and *N*-benzyloxycarbonyl-L-phenylglycine (ionization mechanism), in chloroform solutions, the times for 50 per cent loss of optical rotation are 41, infinity and 800 minutes respectively.

All three mechanisms are ionic and they seem to occur more readily in polar solvents. However, rate of racemization has to be offset against rate of coupling and relatively polar solvents, like *N*,*N*–dimethylformamide, dimethylsulphoxide and tetrahydrofuran, are often ideal for peptide

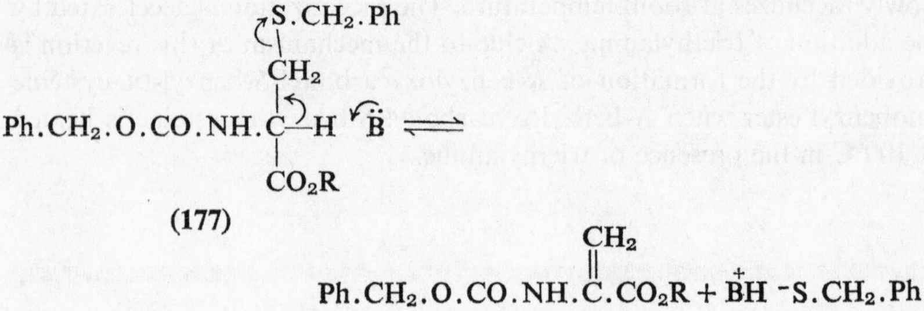

$$Ph.CH_2.O.CO.NH.\overset{\overset{\overset{S.CH_2.Ph}{|}}{CH_2}}{\underset{\underset{CO_2R}{|}}{C}}\text{—}H \;\; \ddot{B} \rightleftharpoons$$

(177)

$$Ph.CH_2.O.CO.NH.\overset{\overset{CH_2}{\|}}{C}.CO_2R + \overset{+}{B}H \; ^{-}S.CH_2.Ph$$

(178)

$$C_6H_4(CO)_2N\text{—}C(=C_6H_5)\text{—}CO_2\text{—}C_6H_4\text{—}NO_2$$

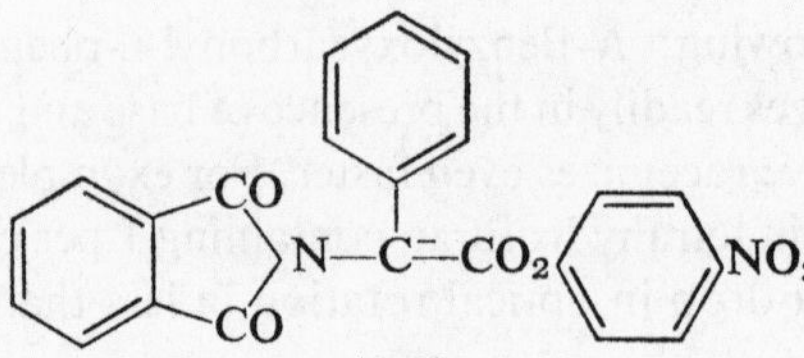

(179)

synthesis. Generally, the presence of salts will tend to facilitate racemization and chloride ions, in particular, due to their basicity in organic solvents, have been faulted from this point of view.

The extent of racemization in synthetic peptides can be estimated by optical measurements on acid hydrolysates of the peptide, but an enzymic method based on the use of an aminopeptidase is much more sensitive and requires less material. Peptides should, of course, be completely degraded by aminopeptidases provided that the terminal amino group is free and the amino acid residues are of the L-configuration. The completeness of the enzymic digestion thus provides a measure of the optical purity of the peptide.

In addition to these general methods, specific peptides have been investigated in which small amounts of diastereoisomers are capable of detection. The synthesis of these peptides has been used as a racemization test for the various coupling methods. Examples of this kind of test are provided by the synthesis of (a) *N*-benzoylleucylglycine ethyl ester from *N*-benzoyl-L-leucine and glycine ethyl ester; (b) *N*-benzyloxycarbonylglycylphenylalanylglycine ethyl ester from *N*-benzyloxycarbonylglycyl-L-phenylalanine and glycine ethyl ester; (c) *N*-trifluoroacetylvalylvaline methyl ester from *N*-trifluoroacetyl-L-valine and L-valine methyl ester; and (d) glycylalanylleucine, after hydrogenolysis, from *N*-benzyloxycarbonylglycyl-L-alanine and L-leucine benzyl ester. In each of these examples, oxazolone formation from the activated residue is possible. The separation and estimation of the optical isomers are achieved, in the first two examples, by recrystallization; in the third, by gas–liquid chromatography; and in the fourth, by ion exchange chromatography. In another ingenious test, *N*-acetyl-L-isoleucine is coupled to glycine ethyl ester and, after hydrolysing the resulting peptide, the ratio of isoleucine (representing retention of the symmetry at the α-carbon atom) to alloisoleucine (representing inversion of the symmetry at the β-carbon atom) is measured by amino acid analysis.

Tests of the above kind have revealed that there is a danger of slight racemization with most of the coupling methods discussed. The azide method was for a long time exceptional in that no racemization could be attributed to it. Now, it seems that other methods of coupling, particularly those involving catechol-type esters, may be racemization-free. Couplings involving *N,N'*-dicyclohexylcarbodiimide proceed without racemization even when the carboxyl component is an acylamino acid or a peptide, provided excess *N*-hydroxysuccinimide is present. It seems that N-hydroxysuccinimide is a better nucleophile than the amide carbonyl so that the *O*-acyl isourea forms an active ester rather than an oxazolone. However, when *N*-hydroxysuccinimide is not present, an appreciable amount of racemization often occurs. In fact, *N,N'*-dicyclohexylcarbodiimide is a very good reagent to use for the synthesis of oxazolones. In all peptide synthesis, strict attention to experimental detail is essential to keep racemization to a minimum.

---

*Exercise 4.18**

Using only mixed anhydrides or *N,N'*-dicyclohexylcarbodiimide for the coupling steps, which of the following routes to the hexapeptide alanylglycylvalylprolylalanylalanine would you not choose if you wanted a product of high optical purity?

(a) Ala. Gly. Val + Pro. Ala. Ala.
(b) Ala. Gly + Val. Pro then + Ala. Ala.
(c) (i) Pro + Ala. Ala. (ii) Val + Pro. Ala. Ala. (iii) Gly + Val. Pro. Ala. Ala. (iv) Leu + Gly. Val. Pro. Ala. Ala.

Strategy of synthesis

When optical purity is important, the limitations of the available coupling methods often dictate the strategy of peptide synthesis. One way in which a racemate-free peptide can be prepared is to start with the *C*-terminal residue of the peptide, say in the form of an ester, and to add to it the other residues as urethane derivatives, for example *N*-benzyloxycarbonyl derivatives, in a stepwise manner (Exercise 4.18(c)*). Another approach involves the preparation of suitably protected smaller peptides which can be joined together subsequently to make the parent peptide. In this 'fragment-condensation' approach, the fragments are chosen so that their *C*-terminal residues either cannot (e.g. glycine) or will not (e.g. proline) racemize under the conditions of the coupling reaction (Exercise 4.18(b)*). If it proves necessary to couple a fragment which possesses a *C*-terminal residue which can racemize, a method known to be racemization-free, for example, the azide method, is employed (Exercise 4.18(a)*).

Peptide synthesis without isolation of intermediates

It will be apparent that the synthesis of peptides is a complicated process. Even the preparation of a simple dipeptide requiring no side-chain protection involves at least four separate steps; larger and more complex peptides can call for a prodigious number of stages and the synthesis of a protein by this approach is an enormous undertaking. In these circumstances, it is understandable that attempts have been made to forego the classical luxury of characterizing intermediates and, since peptide synthesis is essentially a repetitious process, this can be done extensively. Neglecting to isolate the intermediates does not mean that they cannot be studied; their purity can be confirmed by chromatography and, since amino acid analysis is so sensitive, their compositions can be checked without prohibitive loss of material.

Two examples of this type of approach are considered below. Both involve a stepwise construction of the peptide; the first technique employs polymeric materials in conjunction with the synthetic methods already considered; the second approach involves a new reaction.

Use of a polymeric support (*solid-phase synthesis*).

In this approach, the amino acid which is to form the *C*-terminal residue of the peptide is attached to an insoluble high molecular weight polymer and the required peptide is built up on it in a stepwise manner. It is therefore an easy matter to separate the peptide material from small molecular weight impurities and this makes it practicable, at each coupling, to use a large excess of the required amino acid derivative and of any other necessary reagents. By this means, virtually 100 per cent reaction of the peptide can be obtained at each step.

The original and best tried method of this type uses polystyrene beads as the support. Attachment of the *C*-terminal amino acid to the polymer is via a benzyl ester-type bond [as in (**180**)], formed by reacting a salt of the *N*-protected *C*-terminal amino acid with the partially chloromethylated polymer (**181**).

CH_2Cl

$Me_3C.O.CO.NH.CHR^n.CO_2^-$

(**181**)

$CH_2.O.CO.CHR^n.NH.CO.O.CMe_3$

(P) = polymer

(**180**)

Removal of the *N*-protecting group is followed by the addition of the next amino acid residue, usually as its *N*-*t*-butoxycarbonyl derivative. *N,N'*-Dicyclohexylcarbodiimide is generally used as the coupling reagent and *N,N*-dimethylformamide as the solvent. The *t*-butoxycarbonyl groups are cleaved by treating the *N,N*-dimethylformamide-washed resin-bound peptide with cold trifluoroacetic acid. (**180** → **183** etc.).

$CH_2.O.CO.CHR^n.NH_3^+$

(**180**) $\xrightarrow{CF_3.CO_2H}$ (**182**)

(i) base
(ii) $Me_3C.O.CO.NH.CHR^{n-1}.CO_2H$
$C_6H_{11}-N{=}C{=}N-C_6H_{11}$

(**182**)

$CH_2.O.CO.CHR^n.NH.CO.CHR^{n-1}.NH.CO.O.CMe_3$

→ etc.

(**183**)

At the end of the synthesis the peptide is cleaved from the polymer by the use of hydrogen bromide in acetic or trifluoroacetic acids, i.e. cleavage of a benzyl ester.

This 'solid-phase' approach to peptide synthesis has been very successful in specific instances, and it has been fully automated.

Exercise 4.19

Some limitations to the 'solid-phase' approach, as it is described above, will be apparent immediately; further drawbacks and possible side-reactions will become clear on reflection. List as many of these possibilities as you can. You may also care to speculate how the difficulties, assuming that they are real, can be avoided.

Use of N-*carboxyanhydrides*

Although the synthesis of proteins has only recently become a real possibility, high molecular weight polymers composed of amino acids joined together by peptide-type linkages have been available for many years and have proved invaluable as model proteins. These substances are referred to as poly-α-amino acids to distinguish them from peptides and they differ from peptides in two important features. First, the products of a polymerization reaction can rarely be separated completely and so even the best poly-α-amino acid preparation consists of a closely related group of polymers which possess different molecular weights; second, only a limited control of the sequence of the amino acid residues in the polymer is possible.

All of the fundamental coupling techniques discussed above have been used to achieve polymerization, but the most successful polymerization technique has only recently become important in the synthesis of peptides. This technique is based on the use of Leuch's anhydrides, a class of compounds which was discovered over sixty years ago.

*Exercise 4.20**

A suspension of phenylalanine in dioxan reacts with phosgene to give a clear solution which on careful evaporation gives a crystalline solid, A, $C_{10}H_9NO_3$; glycine gives a similar product, B, $C_3H_3NO_3$. When the glycine reaction is interrupted by evaporating the solvent and phosgene, and aniline is added to the mixture, phenylhydantoic acid is produced ($PhNHCONHCH_2CO_2H$). The products, A and B, decompose readily on treatment with base with loss of carbon dioxide. Trace the course of these reactions.

Leuch's anhydrides or *N*-carboxyanhydrides (**184**), as they are more systematically called, are produced by the action of phosgene on the parent amino acid, or phosphorous tribromide on the *N*-benzyloxy-carbonyl amino acid. A variety of initiators, including amines and some metallic salts, for example lithium chloride, cause the anhydrides to polymerize with the evolution of carbon dioxide and the production of poly-α-amino acids, often of very high molecular weight. The mechanism of the polymerization, which is complex, is outside the scope of this book, but it is interesting to note that two types of rate-determining initiation step can seem to be involved. The first is the attack by an amine initiator on the carbonyl of the *N*-carboxyanhydride (**184** → **185**), followed by proton transfer and decarboxylation (**185** → **187**); the second is a proton-withdrawal mechanism (**184** → **188** etc.). Primary and secondary amine initiators seem to act predominantly by the first mechanism, although ionization can be important when the amines, for example diisopropyl-

HN(CHR)CO–O–CO (**184**) $\xrightarrow{HNR^1R^2}$ $HN(CO_2^-)$–CHR–$CO.\overset{+}{N}HR^1R^2$ (**185**) ⟶ $HN(CO_2H)$–CHR–$CO.NR^1R^2$ (**186**)

(**186**) ⟶ $CO_2 + NH_2.CHR.CO.NR^1R^2$ (**187**)

(**184**) $\xrightarrow{B}$ (**188**) $\xrightarrow{(+184)}$ $HN(CO_2^-)$–CHR–CON(CHR)CO–O–CO $\xrightarrow{(+184)}$

$H_2N.CHR.CO.N$(CHR)CO–O–CO + (**188**) + CO_2

amine, are sterically hindered. Tertiary amine initiation takes place via the ionization mechanism. Lithium chloride initiation is a similar anion initiated process.

By the use of efficiently buffered solutions at pH 10·2, it has now proved possible to control the reactivity of *N*-carboxyanhydrides so that the reaction with amino acids can be arrested at the carbamate stage (**189**).

$$\begin{array}{c} R^{n-1} \\ | \\ CH \\ HN \quad CO \;\; :N(H).CHR^{n}.CO_2^- \\ | \qquad | \\ OC\!-\!\!-\!O \;\; H \end{array} \dashrightarrow \begin{array}{c} R^{n-1} \\ | \\ CH \\ NH \quad CO.NH.CHR^{n}.CO_2^- \\ | \\ CO_2^- \end{array}$$

(189)

The *N*-carboxyanhydride may therefore be regarded as an *N*-protected, carboxyl group-activated form of the amino acid. Acidification of the carbamate derivative of the dipeptide leads to the decarboxylation of the resulting carbamic acid (**190** → **191**) and the terminal amino group, thereby exposed, is available for the addition of further residues.

$$(\mathbf{189}) \xrightarrow{H^+} \left[\begin{array}{c} R^{n-1} \\ | \\ CH \\ NH \quad CO.NH.CHR^{n}.CO_2H \\ | \\ CO_2H \end{array}\right] \xrightarrow[CO_2]{\text{bubble } N_2 \text{ through mixture to remove}}$$

(190)

$$\overset{+}{N}H_3.CHR^{n-1}.CO.NH.CHR^{n}.CO_2H \quad (+CO_2) \xrightarrow[(2)\; +\; \begin{array}{c} CHR^{n-2} \\ NH \quad CO \\ CO\!-\!O \end{array}]{(1)\ \text{adjust to buffered pH 10·2}}$$

(191)

$$\begin{array}{c} R^{n-2} \\ | \\ CH \\ NH \quad CO.NH.CHR^{n-1}.CO.NH.CHR^{n}.CO_2^- \\ | \\ CO_2^- \end{array} \xrightarrow{\text{etc.}} \text{required peptide}$$

This sequence of reactions, like *N*-carboxyanhydride polymerization, proceeds with complete retention of optical purity, and has proved very effective. Even so, because of the cumulative difficulties created by side reactions, it seems best not to join more than four or five residues together without isolating the products.

*Exercise 4.21**

When the *N*-carboxyanhydride derived from glycine was reacted in the above manner with phenylalanine, a dibasic acid, $C_{12}H_{14}N_2O_5$ accounted for more than a fifth by weight of the products. Deduce a structure for this compound and show how it might have arisen.

*Exercise 4.22**

Histidine *N*-carboxyanhydride readily undergoes an intramolecular rearrangement by a similar type of mechanism to give a monobasic acid, $C_7H_7N_3O_3$. Formulate the steps involved in the reaction. Can you propose how the original *N*-carboxyanhydride structure might be modified to reduce the likelihood of this type of side-reaction?

The most important side-reaction of the *N*-carboxyanhydride method arises from the premature loss of carbon dioxide from the carbamate. This exposes the amino group of the intermediate peptide for further reaction and represents a breakdown in its protection. In addition, some of the amino component which is meant to react is converted to the carbamate and is therefore blocked. Thus, up to 1 per cent of the starting material always remains after these couplings no matter how much *N*-carboxyanhydride is employed. Thiocarbamates are more stable than carbamates and the use of *N*-thiocarboxyanhydrides (i.e. 2,5-dioxothiazolidines, **192**) has therefore been investigated. Reaction is much more complete in this case (pH 9·5), but racemization does occur to some extent and detracts from the value of these derivatives.

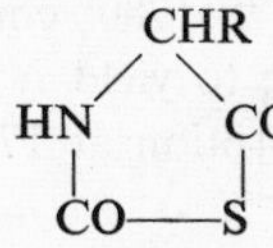

```
      CHR
     /   \
   HN     CO
    |      |
    CO----S
```

(192)

In the case of glycine *N*-carboxyanhydride, formation of an isocyanate (**193**) under the basic conditions of the coupling can be a troublesome side reaction. The isocyanate is, of course, susceptible to attack in its own right and appreciable amounts of substituted ureas (**194**) sometimes result.

$$\text{(glycine NCA)} \rightleftharpoons \underset{\textbf{(193)}}{O{=}C{=}N{-}CH_2{-}CO_2^-} \xrightarrow{RNH_2} \underset{\textbf{(194)}}{R.NH{-}CO{-}NH{-}CH_2CO_2^-}$$

Histidine *N*-carboxyanhydride can form the bicyclic heterocyclic (**195**) by a similar series of reactions.

(**195**)

N-Thiocarboxyanhydrides have a definite advantage in this respect, because the equilibrium between the anhydride and the isocyanate is very much in favour of the anhydride.

Supplementary exercises

Exercise 4.23

N^{α}-[*p*-Nitrobenzyloxycarbonyl]-histidine reacts with *N,N′*-diisopropylcarbodiimide in dioxan solution to give a crystalline solid (A), m.p. = 186–187°, $C_{14}H_{12}N_4O_5$, ν_{max} = 1775–1780 cm^{-1}. This compound reacts quantitatively with benzylamine to yield N^{α}-[*p*-nitrobenzyloxycarbonyl]-histidine benzylamide (no absorption at 1775–1780 cm^{-1}). Explain this sequence of reactions.

Exercise 4.24

N-Benzyloxycarbonylasparagine is readily dehydrated by *N,N'*-dicyclohexylcarbodiimide to the *β*-nitrile (see p. 93). The dehydration does not take place in asparaginyl peptides or esters. When the oxygen of the carboxyl group of *N*-benzyloxycarbonylasparagine is isotopically labelled, i.e. $PhCH_2OCONHCH(CH_2CONH_2)CO^{18}OH$, the label appears in the *N,N'*-dicyclohexylurea. Propose a mechanism for this reaction.

Exercise 4.25

N-Tritylserine reacts with *N,N'*-diisopropylcarbodiimide in an inert solvent to give a compound, m.p. = 193–194°, $C_{22}H_{19}NO_2$, ν_{KBr} = 1820 cm^{-1}, no absorption in the 3300 cm^{-1} region. When treated with benzylamine, this compound is converted quantitatively to the *N*-substituted benzylamide of *N*-tritylserine. Trace the course of these reactions.

Exercise 4.26

No N^{α}-benzyloxycarbonyl-N^{ω}-nitroarginine *p*-nitrophenyl ester could be obtained when the protected amino acid was reacted with *p*-nitrophenol and *N,N'*-dicyclohexylcarbodiimide in the usual way. Instead, a material was isolated which was insoluble in dilute acid or aqueous sodium bicarbonate. It analysed as $C_{14}H_{17}N_5O_5$ [ν_{KBr} = 1730, 1692, 1600 cm^{-1}; λ_{max} = 275 nm, ϵ = 16,120]. Two products were formed when this substance was heated with proline *t*-butyl ester; one of these proved to be nitroguanylproline *t*-butyl ester, the other analysed as $C_{13}H_{16}N_2O_3$ [ν_{KBr} = 1730, 1692 cm^{-1}; no absorption in the 270 nm region]. Explain these observations.

Exercise 4.27

Nitroguanidino derivatives (see Exercise 4.26) can be prepared by reacting an amine with 3,5-dimethylnitroguanylpyrazole. Devise a synthesis of this reagent and indicate how it reacts with amines.

MeC Me

N——N

$O_2N.N{=}C$

NH_2

Note added in proof

It has recently been shown that isotope incorporation does not occur when *N*-benzyloxycarbonyl-*S*-benzyl-L-cysteine is allowed to racemize under basic conditions in the presence of ^{35}S-benzylmercaptan. Hence, racemization cannot proceed by the elimination–addition route suggested in this chapter. The nature of the racemization process in this particular case is, at present, obscure.

CHAPTER FIVE

Natural Peptides: Structure and Synthesis I

So many peptides occur in nature that it would be difficult, even in an appreciably larger volume than the present, to give a full account of them. Here, a few examples have been selected to illuminate points of structural determination or synthesis and to provide an indication of the types of molecule found. This chapter is concerned with what may be regarded as 'normal' peptides, i.e. those which possess the types of structure encountered in proteins. Only the common α-amino acids are involved and they are all of the L-configuration, joined with one exception, by α-peptide bonds.

Glutathione

*Exercise 5.1**

Glutathione, which was first isolated from yeast, exists in both oxidized, $C_{20}H_{32}N_6O_{12}S_2$, and reduced, $C_{10}H_{17}N_3O_6S$, forms. Titration of the reduced form reveals the presence of four ionizable groups, pK_a^1, 2·1; pK_a^2, 3·5; pK_a^3, 8·7; pK_a^4, 9·1. Glutathione gives a positive biuret test and, the reduced form, a purple colour with sodium nitroprusside.

(a) When heated in aqueous solution, reduced glutathione decomposes to two crystalline products, A, $C_5H_7NO_3$, and B, $C_5H_8N_2O_2S$.

Oxidized glutathione, under these conditions, forms A and C, $C_{10}H_{14}N_4O_4S_2$. Hydrolysis gives glutamic acid from A; glycine and cysteine from B; and glycine and cystine from C.

(b) Hydroxyglutaric acid is formed when glutathione is first treated with nitrous acid and then hydrolysed with hydrochloric acid.

(c) A bisthiohydantoin derivative can be isolated when reduced glutathione is heated with ammonium thiocyanate in acetic acid–acetic anhydride solution. Treatment with benzaldehyde followed by hydrolysis gives benzalthiohydantoin.

(d) The ethyl ester of glutathione reacts with 2 moles of phenylmagnesium bromide in the usual way. Acidic hydrolysis of the product gives diphenylacetaldehyde.

Deduce the structure of glutathione and account for these observations.

The tripeptide, glutathione, is found in all living cells and it occurs in particularly high concentration in the lens. A motley assortment of information (Exercise 5.1*) shows that its structure is γ-glutamylcysteinylglycine (**196**; R = H), or the corresponding symmetrical disulphide. Glutathione was first isolated in 1921 and synthesized in 1935 and, in part, this is why the structural evidence is so assorted. With chromatography and other modern techniques, the elucidation of its primary structure would probably have been a routine and straightforward matter. However, rearrangements also contribute to the complexity. Thus, autolysis with the formation of pyrrolidone-2-carboxylic acid (**197**) and the 2,5-dioxopiperazine (**198**) occurs readily when either the reduced or oxidized forms are heated in water (Exercise 5.1(a)*).

$NH.CH(CH_2.S.R).CO.NH.CH_2.CO_2H$
$OC{-}CH_2$
CH_2
$H_2N{-}CH$
CO_2H
(**196**)

⟶

CH_2
OC CH_2
$HN{-}CH{-}CO_2H$
(**197**)
+
$CH_2.S.R$
CH
HN CO
OC NH
CH_2
(**198**)

Evidence for the presence of glutamic acid in the *N*-terminal position is provided by its deamination (Exercise 5.1(b)*) and for glycine in the *C*-terminal position, by its conversion into a thiohydantoin derivative (Exercise 5.1(c)*). The formation of the bisthiohydantoin derivative (**199**) from reduced glutathione (Exercise 5.1(c)*) indicates that it is the γ-carboxyl group of the glutamic acid which is involved in the peptide bond.

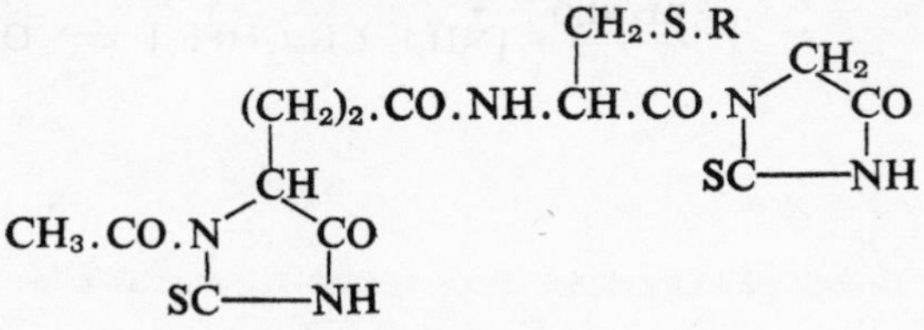

```
                          CH2.S.R
                          |           CH2
           (CH2)2.CO.NH.CH.CO.N     CO
           |                      |      |
           CH                     SC----NH
         /    \
CH3.CO.N      CO
        |      |
        SC----NH
```

(199)

The formulation of glutathione as a γ-glutamyl peptide is, of course, implied in the explanation given for its ready autolysis. Additional evidence is provided by the titration curve; pK_a^1 represents the dissociation of a relatively strong acid whereas pK_a^2 is more like the value obtained for terminal carboxyl groups in peptides. Hence pK_a^1 represents the dissociation of the α-carboxyl group of the terminal glutamic acid residue because, in this situation, the acidity of the carboxyl group is enhanced by the charged α-amino group. It is by no means easy to assign the other pK_a values quoted. *A priori*, either of them could apply to the α-amino or thiol groups. It can be appreciated that the difficulty of identifying ionizing groups can be even more serious when larger, more complex peptides are involved.

Table 5.1 summarizes the accepted pK_a values of the proton-binding groups encountered in proteins. These values tend to be modified when interactions occur between different groups.

Table 5.1. Intrinsic dissociation constants of titratable groups in proteins in dilute aqueous solutions at 25°C (after C. Tanford, *Adv. Protein Chem.* (1968) **23**, 147)

Group	pK_a	Group	pK_a
Terminal COOH	3·6	Thiol	9·0
Aspartyl-β-COOH	4·0	Phenol	9·6
Glutamyl-γ-COOH	4·5	Lysyl-ϵ-NH_3^+	10·4
Imidazole	6·4	Guanidino	12·5
Terminal NH_3^+	7·8		

Further confirmation that glycine is the *C*-terminal residue in glutathione is provided by the formation of diphenylacetaldehyde (**200**) when the ester of the peptide is treated with phenylmagnesium bromide.

$$\mathrm{R.CO.NH.CH_2.CO_2Et \xrightarrow{2PhMgBr} R.CO.NH.CH_2.C(OH)Ph_2}$$

$$\mathrm{\xrightarrow{H^+/H_2O} [\overset{+}{N}H_2{=}CH.CHPh_2] \rightleftharpoons OCH.CHPh_2}$$

(200)

---

*Exercise 5.2**

Outline a potential synthetic route to glutathione.

Exercise 5.3

Is it likely that glutathione could be synthesised by the following route, using *N,N'*-dicyclohexylcarbodiimide as the coupling reagent?

$$\mathrm{Tri.\overset{Tri}{\overset{|}{Cys}} + Gly.OEt \longrightarrow Tri.\overset{Tri}{\overset{|}{Cys}}.Gly.OEt \xrightarrow{(a)} \overset{Tri}{\overset{|}{Cys}}.Gly.OEt \xrightarrow{Tri.Glu}}$$

$$\mathrm{\underset{Tri.Glu}{\llcorner}\overset{Tri}{\overset{|}{-Cys}}.Gly.OEt \xrightarrow{(b)} Glutathione}$$

Glutathione has been synthesized in a dozen or so different ways. Most routes start with the synthesis of a suitably protected cysteinylglycine derivative (**201**) to which a glutamic acid derivative is subsequently coupled (Scheme 5.1).

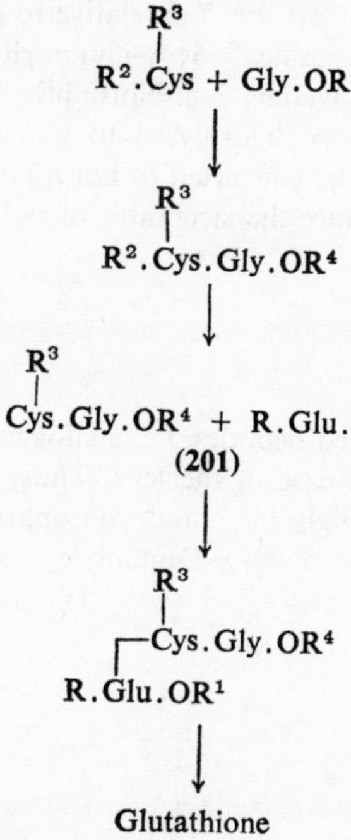

$$R^2.\overset{\overset{R^3}{|}}{Cys} + Gly.OR^4$$

$$\downarrow$$

$$R^2.\overset{\overset{R^3}{|}}{Cys}.Gly.OR^4$$

$$\downarrow$$

$$\overset{\overset{R^3}{|}}{Cys}.Gly.OR^4 + R.Glu.OR^1$$

(201)

$$\downarrow$$

$$R.Glu.OR^1 \text{ — (}\gamma\text{)—} \overset{\overset{R^3}{|}}{Cys}.Gly.OR^4$$

$$\downarrow$$

Glutathione

Scheme 5.1. Generalized pattern of glutathione syntheses

The main difficulties in the synthesis are to protect the thiol adequately and to achieve selective coupling through the γ-carboxyl group of glutamic acid. Thiol protection has most often been accomplished by the *S*-benzyl group, although other protecting groups (e.g. Exercise 5.3) and even the free disulphide have been employed. Coupling at the γ-carboxyl group has been achieved by its selective activation, as in *N*-benzyloxycarbonyl-glutamic acid γ-azide, by selective protection of the α-carboxy group, as in α-ethyl *N*-benzyloxycarbonylglutamate, and by other methods (e.g. Exercise 5.3).

---

*Exercise 5.4**

Ophthalmic acid, another constituent of the lens, can be hydrolysed to give one residue each of glutamic acid, glycine and α-aminobutanoic acid. Glycine is *C*-terminal and glutamic acid, *N*-terminal. The electrophoretic mobility of this peptide at pH 4 is 7·2 relative to glutathione (disulphide form), 7·6; α-glutamyltyrosine 3·3; γ-glutamyltyrosine, 6·9; α-glutamylvaline, 2·8; γ-glutamylvaline, 7·5. Norophthalmic acid, which is also obtained from lens, can be hydrolysed to glutamic acid, glycine and alanine. Glutathione can be converted to norophthalmic acid by treating it with Raney nickel. Deduce the structures of ophthalmic acid and norophthalmic acid.

Groups of closely related peptides, i.e. analogues, are frequently found together and such is the case in the lens. Thus, ophthalmic acid, is γ-glutamyl-α-aminobutanoylglycine, and norophthalmic acid, γ-glutamylalanylglycine (Exercise 5.4*). γ-Glutamylcysteinyl-β-alanine has been isolated from plants.

Oxytocin and vasopressin

---

*Exercise 5.5**

Thermoelectric osmometry indicates that the pituitary hormone, oxytocin, has a molecular weight of 1029 ± 6. Total hydrolysis gives equimolar amounts of isoleucine, cystine, glycine, tyrosine, glutamic acid, aspartic acid, proline and leucine and three times as much ammonia. Reacting the hormone with 2,4-dinitrofluorobenzene prior to hydrolysis results in the loss of half of the cystine and of an indeterminate amount of tyrosine.

When oxytocin is treated with bromine water, two peptides are produced, one of which contains cysteic acid (p. 24) and dibromotyrosine, and the other, cysteic acid and the other remaining amino acid residues. Only one product results when oxytocin is treated with performic acid. Treatment with Raney nickel gives another single product. The following peptides can be obtained by partial hydrolysis of the performic acid-oxidized material: Ile.Glu; Leu.Gly; $CySO_3H$(Leu,Pro); Asp.$CySO_3H$; $CySO_3H$.Tyr; Tyr(Ile.Glu); Asp($CySO_3H$,Pro,Leu). [$CySO_3H$ denotes cysteic acid.]

No thiohydantoin derivatives are formed when ammonium thiocyanate-treated oxytocin is hydrolysed, but, if oxytocin is subjected to a brief period of hydrolysis prior to the thiocyanate treatment, subsequent hydrolysis gives detectable amounts of thiohydantoin. Asparagine and glutamine are both present in papain (i.e. enzymic) digests of oxytocin.

Deduce the structure of oxytocin.

Although the presence of cystine in a peptide usually complicates both degradation and synthesis, glutathione offers no difficulties on this account because only one cystine residue is involved and the molecule is symmetrical about the disulphide bond. The pituitary hormone, oxytocin, is a degree more complex (Exercise 5.5*). In this case, there is still only one cystine residue in the molecule, but the element of symmetry is missing. The formation of only one product by performic acid oxidation indicates that the disulphide bond is an intrachain link.

Careful integration of the sequences revealed by partial hydrolysis indicates the structure of oxytocin (**202**). The cleavage of this molecule by bromine water seemed contrary when it was first observed, but, with present-day knowledge of selective chemical cleavage, it is not difficult to visualize that the splitting of the tyrosyl–isoleucine bond is probably related to the *N*-bromosuccinimide cleavage of tyrosyl peptides (p. 32).

```
                    OH
                    |
                  (C6H4)         Et  Me
                    |              \ /
    Cys (1)        CH2             CH
                    |               |
H2N.CH.CO.NH.CH.CO.NH.CH.CO ─────────┐
     |          Tyr (2)      Ile (3)  │
    CH2                               │
     |                                │
     S                                │
     |                                │
     S                       (CH2)2.CO.NH2
     |                                │
    CH2        CH2.CO.NH2       |     │
     |              |           |     │
CO.CH.NH.CO.CH.NH.CO.CH.HN ◄─────────┘
|  Cys (6)      Asn (5)     Gln (4)
|
N—CH.CO.NH.CH.CO.NH.CH2.CO.NH2
/   \           |        Gly (9)
CH2   CH2       CH2
 \   /          |
  CH2          CHMe2
 Pro (7)      Leu (8)
```

(**202**)

Similar degradative studies have shown that the hormone vasopressin, which can be extracted along with oxytocin from the posterior pituitary gland, has the same structure **(202)** except that a phenylalanine residue is in position 3 and a basic residue in position 8. In most species, the basic residue is arginine, but in the pig and hippopotamus, for example, it is lysine. Other analogues which have been isolated, include mesotocin, from amphibians, which has the oxytocin structure except that isoleucine is in position 8; and isotocin, from bony fishes, which resembles mesotocin except that a serine residue is in position 4. The biological properties of both of these compounds are qualitatively similar to those of oxytocin. Vasotocin, from amphibians, has the same structure as arginine–vasopressin except that isoleucine is in position 3. This hormone resembles vasopressin in its biological activity. In all species studied, two dominant hormones, one of which is related to oxytocin and the other to vasopressin, have been found in the posterior pituitary gland. Most peptides, like the pituitary hormones, exhibit species variations.

---

*Exercise 5.6**

Explain what was happening in the following experiments:

Small pieces of sodium were added to a stirred solution of oxytocin in liquid ammonia until the resulting blue colour just persisted. Benzyl chloride, equivalent to the total amount of sodium which had been consumed, was subsequently added to the stirred reaction mixture. Evaporation of the ammonia gave a biologically inert material.

This substance was redissolved in liquid ammonia and sodium was added once more to the stirred mixture until the blue colour just persisted. Acetic acid, equivalent to the total amount of sodium used, was carefully added and the ammonia was allowed to evaporate. The residue was dissolved in water and the acidity of the solution was adjusted to pH 3 by the addition of acetic acid. At first, a drop of the resulting solution gave a purple colour when tested with sodium nitroprusside, but, after air had been bubbled through the solution for an hour or so, this test was negative. The solution was biologically active and almost equivalent in activity to the hormone taken.

---

*Exercise 5.7**

Devise a potential synthetic route to oxytocin.

Since it is symmetrical about the disulphide bond, glutathione can be synthesized from cystinyl peptides; oxytocin cannot be prepared conveniently in this way (e.g. **203** → **204**) because disulphide interchange (e.g. **203** → **205**, **206** → **207**, etc.) is likely to occur.

A.Cys.B | A.Cys.B + C.Cys.D | C.Cys.D
(205)

A.Cys.Tyr | A.Cys.Tyr + C.Cys.D | C.Cys.D
(207)

A.Cys.B | C.Cys.D **(203)** ⟶ A.Cys.Tyr | C.Cys.D **(206)** ---→ A.Cys.Tyr.Ile.Gln.Asn | C.Cys.D ⟶

A.Cys.Tyr.Ile | C.Cys.Asn.Gln ⟶ Cys.Tyr.Ile | Cys.Asn.Gln | Pro.Leu.Gly.NH_2
(204)

Fortunately, oxytocin can be reduced reversibly so that the nonapeptide can be built up from suitably protected cysteine residues and converted to the disulphide form in a final oxidative step. The recovery of oxytocin from the reduced (dithiol) form (**208**, R = H) is exemplified in Exercise 5.6*, where the reversible formation of *S,S'*-dibenzyloxytocin (**208**, R = CH_2Ph) is also illustrated. In most syntheses of oxytocin, cysteine has been incorporated as its *S*-benzyl derivative and the final steps have generally parallelled the second part of this exercise fairly closely. However, other cysteine derivatives, including *S*-trityl, *S*-4,4'-dimethoxybenzhydryl and *S*-carbamoylcysteine, have also been employed. Usually, in these cases, the dithiol has been produced from the protected nonapeptide under the appropriate conditions and converted to the disulphide in the way described.

$$\underset{\displaystyle R}{\underset{|}{\text{A.Cys}}}\text{.Tyr.Ile.Gln.Asn.}\underset{\displaystyle R}{\underset{|}{\text{Cys}}}\text{.Pro.Leu.Gly.NH}_2$$

(208)

---

*Exercise 5.8**

The yield of oxytocin obtained by aerial oxidation of the dithiol in aqueous solution is concentration-dependent. At high thiol concentrations the yield of oxytocin tends to be relatively poor. Explain these observations.

Aerial oxidation of the dithiol gives variable yields of oxytocin and, to some extent, the yield is concentration-dependent, particularly at high thiol concentrations when dimer formation is a serious side reaction. An improved technique for the conversion of the dithiol to the disulphide form involves the use of 1,2-diiodoethane (**209** → **210**).

$$\underset{\textbf{(209)}}{\text{R—S}^-} + \text{I.CH}_2\text{.CH}_2\text{.I} \longrightarrow \text{R.S.I} + \text{CH}_2\text{=CH}_2 + \text{I}^-$$

$$\text{R—S—I} + \text{RS}^- \longrightarrow \underset{\textbf{(210)}}{\text{R.S.S.R}} + \text{I}^-$$

The *S,S′*-ditrityl nonapeptide can be converted directly to the disulphide by a reaction with iodine in acetone solution (**211** → **212**).

$$\underset{\textbf{(211)}}{\text{R—S—Tri}} \xrightarrow{\text{I}_2/\text{MeOH}} \text{R—S—I} + \text{Tri.OMe} + \text{HI}$$

$$\text{R—S—I} \xrightarrow{\text{RSI}} \underset{\textbf{(212)}}{\text{R.S.S.R}} + \text{I}_2$$

$$\text{R—S—I} \xrightarrow{\text{R.S.Tri/MeOH}} \underset{\textbf{(212)}}{\text{R.S.S.R}} + \text{Tri.OMe} + \text{HI}$$

Numerous individual approaches to the synthesis of the protected *S,S′*-dibenzyl nonapeptide (**208**; R = CH_2Ph) have been reported. One of classical importance, in that it emphasized for the first time the importance of stepwise synthesis, is illustrated in Scheme 5.2. *N*-Benzyloxycarbonyl-amino acid *p*-nitrophenyl esters were employed in this synthesis and, at each stage prior to the formation of the protected nonapeptide, the benzyloxycarbonyl group was cleaved with hydrogen bromide in acetic acid. Finally, the fully-protected nonapeptide was reduced in sodium–liquid ammonia and oxytocic activity was generated in the manner described above. More recently, oxytocin has been synthesized by the solid-phase approach via the same protected nonapeptide.

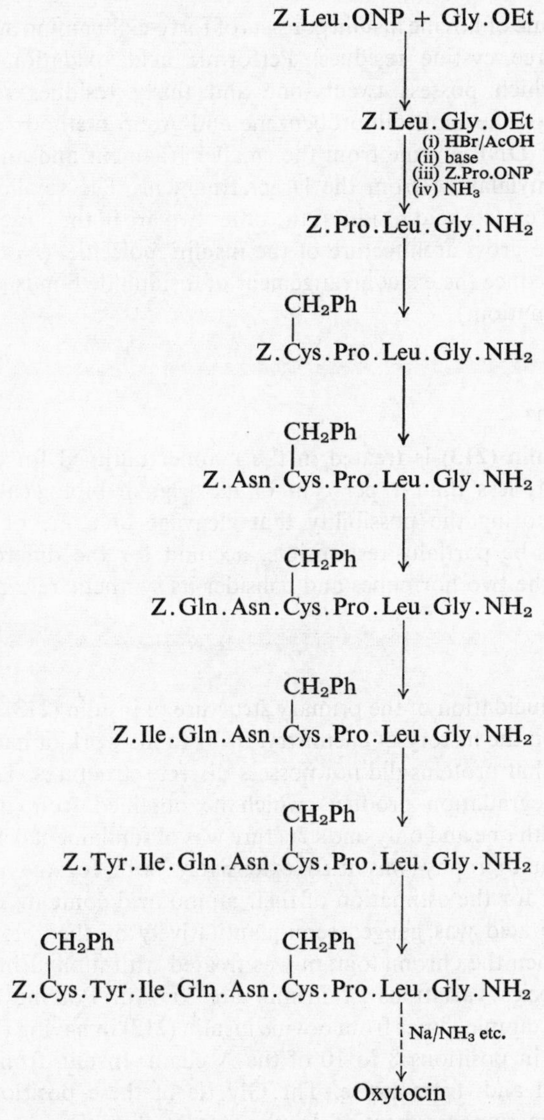

Scheme 5.2. Stepwise synthesis of oxytocin (Z = $PhCH_2OCO$; ONP = $OC_6H_4NO_2p$)

Insulin

*Exercise 5.9**

The molecule of bovine insulin consists of forty-eight amino acid residues including three cystine residues. Performic acid oxidation gives two fragments which possess twenty-one and thirty residues respectively. Application of the dinitrofluorobenzene end-group method produces an equivalent of DNP-glycine from the smaller fragment and an equivalent of DNP-phenylalanine from the larger fragment. The smaller fragment includes four cysteic acid residues; the other two are in the larger fragment.

Deduce the gross architecture of the insulin molecule. (Note: It is not possible to deduce the exact arrangement of disulphide bonds on the basis of this information.)

*Exercise 5.10**

When insulin (**213**) is treated in the manner outlined for oxytocin in Exercise 5.6*, less than 1 per cent of its original biological activity is restored. Ignoring the possibility that cleavage of some of the amide bonds might be partially responsible, account for the difference in the recovery of the two hormones and consider its synthetic relevance.

Sanger's elucidation of the primary structure of insulin (**213**) will remain a landmark in the history of chemistry. Prior to his work, it had even been maintained that proteins did not possess discrete structures. The fact that all of the degradation products which he obtained from insulin were consistent with one and only one structure was of fundamental significance. Paper chromatography was used extensively both for the isolation of peptides and for the estimation of their amino acid content. The amount of an amino acid was gauged semiquantitatively by the colour which it produced when the chromatogram was treated with ninhydrin.

Many species variations of insulin are known. Porcine and canine insulin, for example, differ from bovine insulin (**213**) in having the sequence Thr.Ser.Ile in positions 8 to 10 of the A chain. Insulin from sheep has Ala.Gly.Val and, from horse, Thr.Gly.Ile in these positions. Human insulin is the same as porcine insulin except that threonine is the *C*-terminal residue of the B chain instead of alanine.

1 2 3 4 5 6 7 8 9 10 11 12 13 14 15

A. Gly.Ile.Val.Glu.Gln.Cys.Cys.Ala.Ser.Val.Cys.Ser.Leu.Tyr.Gln.

B. Phe.Val.Asn.Gln.His.Leu.Cys.Gly.Ser.His.Leu.Val.Glu.Ala.Leu.

16 17 18 19 20 21 22 23 24 25 26 27 28 29 30

A. Leu.Glu.Asn.Tyr.Cys.Asn

B. Tyr.Leu.Val.Cys.Gly.Glu.Arg.Gly.Phe.Phe.Tyr.Thr.Pro.Lys.Ala

(213)

In view of the multiplicity of disulphide bonds in the insulin molecule, it is hardly surprising that insulin activity cannot be recovered by simply reoxidizing the reduction products (Exercise 5.10*). An unequivocal synthesis of insulin from cysteine would require the use of three different sulphur protecting groups (**214**; M, N, O) capable of selective removal. At least two of them would need to be removed without the disruption of preformed disulphide bonds (**214** → **215**).

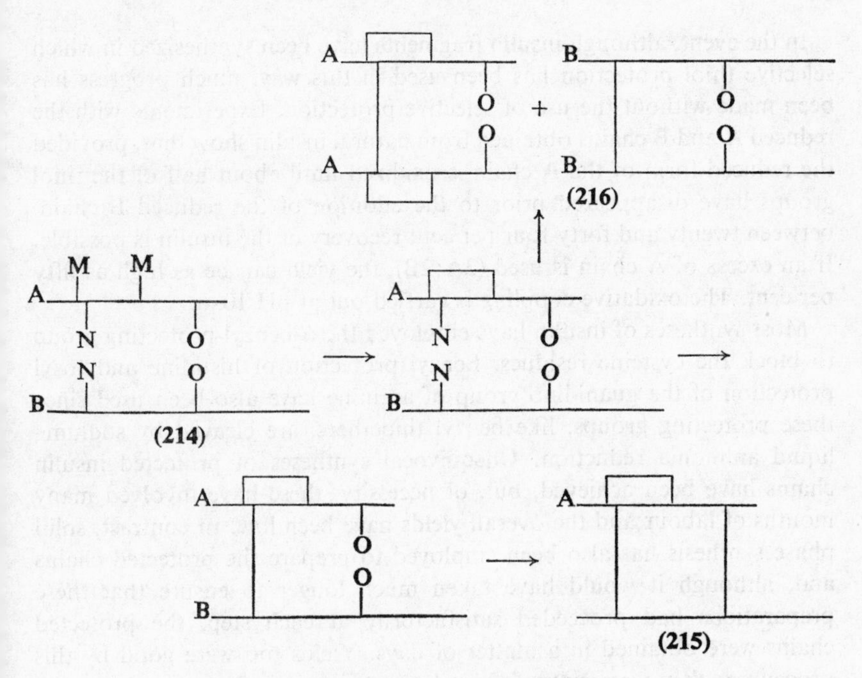

Diagrammatic synthesis of insulin

However, even with distinctive thiol-protecting groups, stepwise cleavage and disulphide bond formation would not mean *per se* that insulin would result. For example, the symmetrical disulphides involving two A chains or two B chains (**216**) could be formed. To avoid this difficulty it has been suggested that the terminal amino groups of the A and B chains should be linked via a polyvalent amino-protecting group (**217**, G) prior to the formation of the first interchain disulphide bridge. The formation of this disulphide would then give a cyclic derivative (**218**), for which, by analogy with the ring system of oxytocin, some degree of stability might be predicted. Finally, when the other disulphide bonds had been formed, the amino-protecting group would be removed.

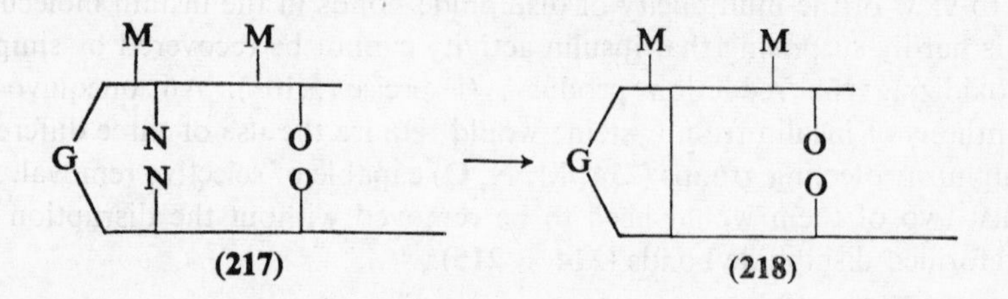

In the event, although insulin fragments have been synthesized in which selective thiol protection has been used in this way, much progress has been made without the use of selective protection. Experiments with the reduced A and B chains obtained from natural insulin show that, provided the reduced form of the A chain is oxidized until about half of the thiol groups have disappeared prior to the addition of the reduced B chain, between twenty and forty-four per cent recovery of the insulin is possible. If an excess of A chain is used (3A:2B), the yield can be as high as fifty per cent. The oxidative coupling is carried out at pH 10·6.

Most syntheses of insulin have employed the *S*-benzyl-protecting group to block the cysteine residues. Benzyl protection of histidine and tosyl protection of the guanidino group of arginine have also been used since these protecting groups, like benzyl thioethers, are cleaved by sodium–liquid ammonia reduction. Unequivocal syntheses of protected insulin chains have been achieved, but, of necessity, these have involved many months of labour and the overall yields have been low. In contrast, solid phase synthesis has also been employed to prepare the protected chains and, although it would have taken much longer to ensure that these preparations had proceeded satisfactorily at each step, the protected chains were obtained in a matter of days. Yields too were good by this procedure; they were sixty-nine and twenty-one per cent overall for the protected A and B chains respectively, based on the amounts of the *C*-terminal amino acids attached to the polymer. These syntheses illustrate the enormous potential of the solid phase approach, even bearing in mind its uncertainties.

Although insulin can be recovered in reasonable yield from reduced insulin by the method outlined above, much lower yields (approximately 1–2 per cent) are obtained when the individual synthetic chains are combined. This is because undesirable side reactions, including de-sulphurization and cleavage of the peptide chain, occur during sodium–liquid ammonia reduction. These side reactions are not important in the preparation of relatively small peptides, but they are so serious in the insulin studies that the value of *S*-benzyl, N^{im}-benzyl and guanidino-tosyl protection in the synthesis of other large peptides is questionable.

Exercise 5.11

Insulin has been partially synthesized by the following route. The *N*-benzyloxycarbonyl hexadecapeptide symmetrical disulphide which possesses the residue sequence of the *N*-terminal sixteen residues of the B chain, was combined by the azide method with the tetradecapeptide symmetrical disulphide equivalent to the sequence of the fourteen *C*-terminal residues of the B chain. *t*-Butyl protecting groups were used on side chains and for the *C*-terminal carboxyl groups; guanidino groups were protonated. The main product of the coupling was a polymer of the 1–30 residue sequence. After appropriate treatment, the synthetic B chain obtained from this product was combined with natural A chain. Outline the synthesis diagrammatically in the style of (**214** → **215**) and consider whether this approach is likely to be better than the *S*-benzyl-cysteine approach discussed above.

Ribonuclease

The enzyme ribonuclease, isolated from bovine pancreas, has a significantly larger molecular weight than insulin. It has a single peptide chain which is composed of one hundred and twenty-four amino acid residues, including four cystine residues. There are therefore four intrachain disulphide bonds. This increase in complexity calls for greater sophistication in the degradative techniques employed for structural investigations and, indeed, the first primary structure proposed for ribonuclease, which was based on the semiquantitative methods of the insulin study, together with applications of the Edman degradation, contained several inaccuracies. The correct structure **(219)** was not obtained until quantitative techniques, which included automated amino acid analyses at every possible step in the investigation, had been introduced.

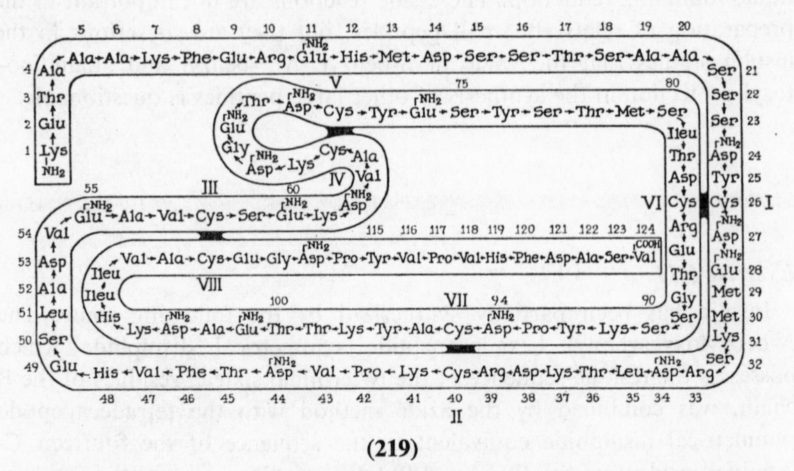

(219)

Amino acid sequence of bovine ribonuclease (from D. G. Smyth, W. H. Stein and S. Moore, *Journal of Biological Chemistry*, 1963, **238**, 228. Reproduced by courtesy of the authors and publishers)

Despite its greater complexity, ribonuclease, so far as the disulphide bonds are concerned, is not such a difficult synthetic proposition as insulin. Reduction of the disulphide bonds of ribonuclease gives a biologically inert product, but most of the biological activity can be recovered by oxidation. Presumably, this implies that the interactions of the individual amino acid residues in reduced ribonuclease hold the peptide chain in the 'right' conformation during oxidation for the disulphides to reform correctly. A preparation which exhibits ribonuclease activity has been synthesized by the 'solid-phase' approach, whilst the *S*-protein of ribonuclease (residue sequence 21–124; see Chapter 7) has been synthesized by the *N*-carboxyanhydride method.

In the solid-phase synthesis, benzyl protection was used for the side chains of cysteine, aspartic and glutamic acids, serine, threonine and tyrosine residues; nitro protection for the guanidino group of arginine;

Table 5.2. Some natural peptides which have been successfully synthesized. All of these peptides possess pharmacological activity and most, perhaps all of them, are hormones.

Name	Origin	Structure
Angiotensin	Plasma (horse and pig)	Asp.Arg.Val.Tyr.Ile.His.Pro.Phe
Bradykinin	Plasma	Arg.Pro.Pro.Gly.Phe.Ser.Pro.Phe.Arg
Kallidin	Plasma	Lys.Arg.Pro.Pro.Gly.Phe.Ser.Pro.Phe.Arg
Methionyllysylbradykinin	Plasma	Met.Lys.Arg.Pro.Pro.Gly.Phe.Ser.Pro.Phe.Arg
Phyllokinin	Amphibian skin	Arg.Pro.Pro.Gly.Phe.Ser.Pro.Phe.Arg.Ile.Tys
α-Melanocyte-stimulating hormone (α-MSH)	Anterior pituitary gland	Ac.Ser.Tyr.Ser.Met.Glu.His.Phe.Arg.Trp.Gly.Lys.Pro.Val.NH_2
β-MSH	Anterior pituitary gland (pig)	Asp.Glu.Gly.Pro.Tyr.Lys.Met.Glu.His.Phe.Arg.Trp.Gly.Ser.Pro.Pro.Lys.Asp
Corticotropin (ACTH)	Anterior pituitary gland	Ser.Tyr.Ser.Met.Glu.His.Phe.Arg.Trp.Glu.Lys.Pro.Val.Gly.Lys.Lys.Arg.Arg.Pro.Val.Lys.Val.Tyr.Pro.Asp.Gly.Ala.Glu.Asp.Glu.Leu.Ala.Glu.Ala.Phe.Pro.Leu.Glu.Phe
Gastrin	Stomach (human)	⌐Glu.Gly.Pro.Trp.Leu.Glu_5.Ala.Tys.Gly.Trp.Met.Asp.Phe.NH_2
Caerulin	Amphibian skin	⌐Glu.Gln.Asp.Tys.Thr.Gly.Trp.Met.Asp.Phe.NH_2
Secretin	Intestine (pig)	His.Ser.Asp.Gly.Thr.Phe.Thr.Ser.Glu.Leu.Ser.Arg.Leu.Arg.Asp.Ser.Ala.Arg.Leu.Gln.Arg.Leu.Leu.Gln.Gly.Leu.Val.NH_2
Glucagon	Pancreas	His.Ser.Gln.Gly.Thr.Phe.Thr.Ser.Asp.Tyr.Ser.Lys.Tyr.Leu.Asp.Ser.Arg.Arg.Ala.Gln.Asp.Phe.Val.Gln.Tyr.Leu.Met.Asn.Thr
Thyrocalcitonin	Thyroid (pig)	Cys.Ser.Asn.Leu.Ser.Thr.Cys.Val.Leu.Ser.Ala.Tyr.Trp.Arg.Asn.Leu.Asn.Asn.Phe.His.Arg.Phe.Ser.Gly.Met.Gly.Phe.Gly.Pro.Glu.Thr.Pro.NH_2 (bracket linking Cys 1 and Cys 7)

Abbreviations: Tys = tyrosine sulphate (ester); Ac = acetyl; ⌐Glu = **pyrrolidone carboxylic acid.**

and benzyloxycarbonyl protection for the N^{ϵ}-amino group of lysine. The entire protected peptide chain was built up on the polymeric support by the automated method. It required 369 chemical reactions and 11,931 steps. Sodium–liquid ammonia was used in the last stages of the synthesis. Unfortunately, even after extensive purification, the final preparation only possessed about thirteen per cent of the expected biological activity and it probably consisted of a mixture of peptides closely related to ribonuclease.

The use of protecting groups was largely avoided in the *N*-carboxyanhydride approach to the *S*-protein. Arginine, glutamic acid and aspartic acid were introduced as the corresponding *N*-carboxyanhydrides; histidine as its *N*-thiocarboxyanhydride; the cysteine residues were blocked by acetamidomethyl groups; and the lysine amino groups by the benzyloxycarbonyl group. Sodium–liquid ammonia treatment was therefore not mandatory and, in fact, liquid hydrogen fluoride was used for the final cleavage step. This synthesis involved the preparation of nineteen relatively small peptides. Forty per cent of the peptide bonds were made via *N*-carboxyanhydrides or *N*-thiocarboxyanhydrides and other residues were introduced as *N*-*t*-butoxycarbonylamino acid hydroxysuccinimide esters. These active esters were used for the introduction of asparagine, serine and threonine and for the introduction of the *N*-terminal residue into each fragment. The azide coupling method was used in the final assembly of the fragments.

These attempts at the synthesis of ribonuclease—albeit inconclusive at present—encourage one to look hopefully ahead to the time when all proteins will be within the scope of synthetic methods. However, it must be remarked that the synthesis of quite small peptides still cannot be embarked upon lightly, even by workers experienced in the field.

Other peptides

Difficulties associated with the presence of cystine in a peptide have received particular emphasis in this chapter, but a great many peptides have been studied which do not contain cystine. Naturally, the absence of cystine from a peptide does not automatically mean that its investigation will be straightforward. The problems encountered are as diverse as the structures themselves. A few representative peptides, selected from the many which have been the subjects of successful structural and synthetic studies, are listed in Table 5.2.

CHAPTER SIX

Natural Peptides: Structure and Synthesis II

Peptides which Possess 'Unusual' Features

For the most part, the peptides considered in the last chapter were composed of L-α-amino acids joined by α-peptide bonds, although a few unusual features, including the γ-glutamyl peptide linkage in glutathione and the terminal acetyl group in α-melanocyte stimulating hormone, were noted. Bacteria and fungi, in particular, are sources of peptides which differ much more fundamentally from proteins. Some of these 'unusual' peptides play a structural role in the organism, whereas the functions of others are obscure. Not infrequently, they have antibiotic properties and are extremely toxic to animals.

*Exercise 6.1**

The capsule of the anthrax bacillus is composed of a protein-like material which gives, on hydrolysis, D-glutamic acid only. Titration with NaOH reveals that there is approximately one dissociable group (COOH) more than the total number of residues, as calculated from the measured molecular weight. Methylation of the material with methanolic hydrogen chloride, followed by borohydride reduction and total hydrolysis, gives δ-hydroxy-γ-aminovaleric acid and a trace of 2-amino-1,5-pentanediol. What is the structure of the capsular material?

*Exercise 6.2**

A basic, carboxypeptidase-resistant peptide, which can be isolated from bacterial cell wall, hydrolyses to give DL-alanine, L-lysine and D-glutamic acid in mole ratios 2:1:1. Aminopeptidase digestion of the peptide liberates 1 mole of L-alanine. Edman degradation is normal for one cycle only, during which 5-methyl-3-phenyl-2-thiohydantoin is liberated. At the next cycle, ammonia is formed and the remaining phenylthiohydantoin is not identical with any of those derived from the common amino acids. Treatment of the residual peptide with 2,4-dinitrofluorobenzene, followed by total hydrolysis, gives DNP-glutamic acid, N^{ϵ}-DNP-lysine and alanine.

Non-protein amino acids and peculiar linkages

The unusual peptides are often distinguished by being composed, either entirely or in part, of D-amino acid residues. Thus, the peptide which constitutes the capsule of the anthrax bacillus (Exercise 6.1*) is formed entirely from D-glutamic acid residues, whilst the structural peptide of the cell wall of *Staphylococcus aureus* (Exercise 6.2*) possesses both D-glutamic acid and D-alanine residues. Sometimes, the peptide bonds encountered are also distinctive. Glutamic acid, for example, in both anthrax and *S.aureus* peptides, is linked through its γ-carboxyl group. The former peptide is poly-γ-D-glutamic acid, and the latter, L-alanyl-D-isoglutaminyl-L-lysyl-D-alanine.

*Exercise 6.3**

An optically inactive substance, $C_7H_{14}N_2O_4$, which can be isolated from acid hydrolysates of a bacterium, evolves two equivalents of carbon dioxide when it is treated with ninhydrin. When treated with nitrosyl bromide it gives a dibromide. Catalytic reduction of the dibromide gives pimelic acid. It is not possible to resolve the starting material into optical antipodes. Propose a structure to account for these observations.

---

*Exercise 6.4**

An optically active compound [$C_9H_{11}NO_2$; infrared, ν_{max} = 1500, 730, 700 cm^{-1}; ultraviolet λ_{max} = 252, 257, 262, 268 nm (low intensity) and high end absorption from 239 nm] can be obtained by very mild acidic hydrolysis of the antibiotic etamycin. More vigorous hydrolysis gives a form with zero optical rotation. The substance is not destroyed by treatment with nitrous acid, but it can be oxidized to methylamine, carbon dioxide and an aldehyde. Assign a structure to it which would explain these findings.

---

*Exercise 6.5**

Another amino acid ($C_8H_{17}NO_2$), which can be obtained by the hydrolysis of etamycin, is optically active and the fact that its rotation becomes more positive on acidification is taken to indicate that it is an L-amino acid. Treatment with nitrosyl chloride gives carbon dioxide, methylamine and an aldehyde. The latter can be shown by analysis, and from the melting point of its 2,4-dinitrophenylhydrazone derivative, to be the same as the aldehyde liberated by the ozonolysis of ergosterol (**220**).

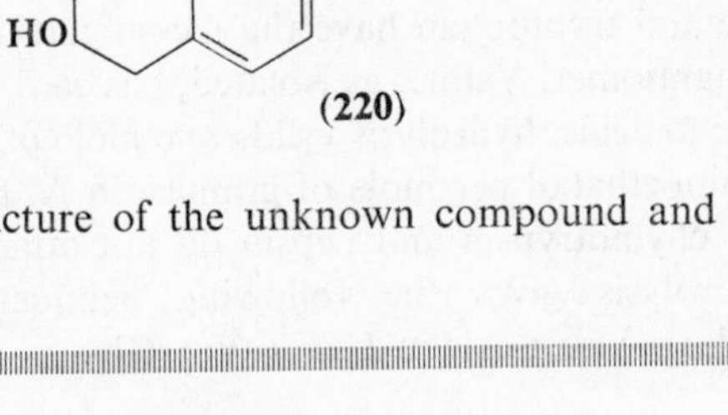

Me
Me
HO

(**220**)

Deduce the structure of the unknown compound and indicate possible ambiguities.

More than seventy amino acids of diverse structures have been found as constituents of antibiotics and related compounds. Examples include *meso*-2,6-diaminopimelic acid (**221**) which occurs instead of L-lysine in the cell walls of gram-negative bacteria (Exercise 6.3*); and the *N*-methylamino acids (**222**) and (**223**) which are obtained from etamycin (Exercises 6.4* and 6.5*).

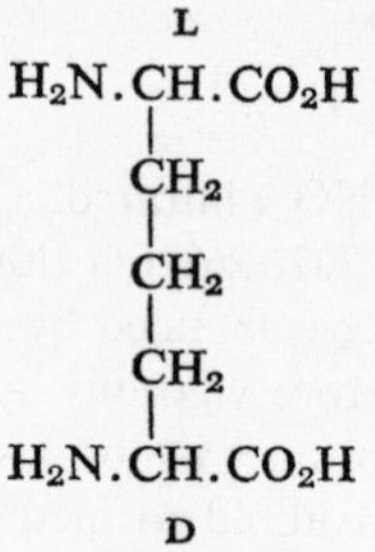

$$\begin{array}{c} \text{L} \\ H_2N.CH.CO_2H \\ | \\ CH_2 \\ | \\ CH_2 \\ | \\ CH_2 \\ | \\ H_2N.CH.CO_2H \\ \text{D} \end{array}$$

(221)

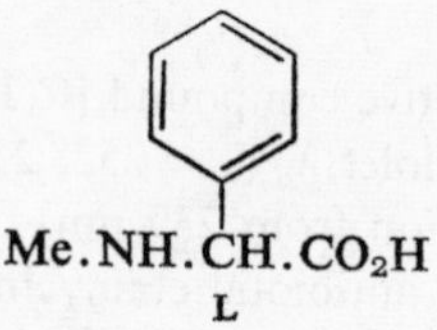

$$C_6H_5 \text{ (phenyl)} - Me.NH.\underset{\text{L}}{CH}.CO_2H$$

(222)

$$\begin{array}{c} Me \quad Me \\ \diagdown\,\diagup \\ CH \\ | \\ CHMe \text{ (stereochemistry not known)} \\ | \\ MeNH.CH.CO_2H \\ \text{L} \end{array}$$

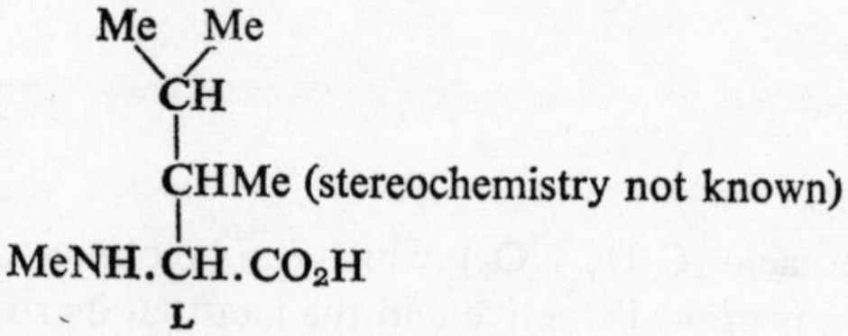

(223)

The bacterial cell wall peptide forms part of a composite structure which involves a polysaccharide component.

---

*Exercise 6.6**

Gramicidin A, a chemically neutral antibiotic produced by *Bacillus brevis*, gives the following amino acid analysis; Gly_1; Ala_2; Val_4; Leu_4; Trp_4. The alanine and tryptophan have the L-configuration, whereas the leucine is the D-enantiomer. Valine, as isolated, has zero optical rotation. In addition to amino acids, hydrolysis yields one mole of formic acid and one mole of 2-aminoethanol per mole of gramicidin A. Enzymes such as nagarse, pronase, chymotrypsin and pepsin do not attack the molecule. Partial acid hydrolysis gives the following peptides: D-Val.L-Val; L-Val.D-Val; L-Ala.D-Val; L-Ala.D-Leu; L-Val.Gly.

The n.m.r. spectrum of gramicidin A in deuteromethanol shows a broad peak at 1·55τ approximately equal to one twelfth to one fourteenth of the indole protons at 4·8–4·0τ. This peak is not present after the antibiotic has been treated with 1·5M hydrogen chloride in methanol for one hour at room temperature. The material so treated is basic and Edman degradation shows it to have the following sequence Val.Gly.Ala.Leu.Ala.Val. Val.Val.Trp.Leu. (Trp.Trp.Trp, Leu,Leu, 2-aminoethanol). Treatment of gramicidin A under appropriate conditions with *N*-bromosuccinimide gives 2-aminoethanol together with complex oxidation products of tryptophan, but no 5-bromodioxindolalanine spirolactone is formed.

Deduce the structure of gramicidin A.

Heteromeric peptides

Many antibiotics possess components other than amino acids. For example, gramicidin A (Exercise 6.6*) differs from 'normal' peptides not only in that it is composed of alternate L- and D-amino acid residues, but also because it possesses *N*-terminal formyl and *C*-terminal ethanolamide moieties (**224**).

HCO.L-Val.Gly.L-Ala.D-Leu.L-Ala.D-Val.L-Val.D-Val.
L-Trp.D-Leu.L-Trp.D-Leu.L-Trp.D-Leu.L-Trp.$NH(CH_2)_2OH$

(**224**)

Peptides which contain components which are not amino acids are called *heteromeric* peptides to distinguish them from *homomeric* peptides which are built up solely of amino acid residues. Twenty or so different carboxylic acids and more than half a dozen amines, heterocyclic compounds and sugars have been found as components of various antibiotics.

---

*Exercise 6.7**

Hydrolysis of the antibiotic gramicidin S gives equimolar proportions of L-leucine, L-ornithine, L-proline, L-valine and D-phenylalanine. The peptide is not deaminated by reaction with ninhydrin but nitrogen is rapidly evolved when it is treated with nitrous acid. Reaction with 2,4-dinitrofluorobenzene gives a yellow substance from which, by total hydrolysis, N^{δ}-DNP ornithine can be obtained. If insufficient dinitrofluorobenzene is employed for the initial reaction to go to completion, two yellow derivatives can be isolated. The extinction of what is presumed to be the mono-DNP derivative indicates its molecular weight to be 1300.

Peptides with the following sequences can be obtained from gramicidin S by partial hydrolysis; Val.Orn; Leu.Phe: Orn.Leu; Pro.Val: Phe.Pro.

Deduce the structure of gramicidin S and consider how this compound might be synthesized.

Homodetic cyclic peptides

Antibiotics generally differ from other peptides in grosser details as well as in the nature of their individual components. Gramicidin A and its near relatives are unique in being the only antibiotics known to possess linear structures. The tyrocidines, which are also obtained from *B. brevis*, and gramicidin S, which is obtained from a similar organism, are both cyclic. Gramicidin S, cyclodi-(L-Val.L-Orn.L-Leu.D-Phe.L-Pro), and the four known variants (A–D) of tyrocidine (**225**; A,X = D-Phe, Y = L.Phe Z = D-Phe; B,X = D-Phe, Y = L-Trp, Z = D-Phe; C,X = D-Trp, Y = L-Trp; Z = D.Phe; D,X = D-Trp, Y = L-Trp; Z = D-Trp.) are closely related to each other. Cyclic peptides in which the rings are composed wholly of amino acids joined by peptidic linkages are called homodetic cyclic peptides. One peptide of this type has been isolated from the leaves of the tree *Evodea xanthoxyloides* (Exercise 2.16) but homodetic cyclic peptides have not been detected in other higher organisms.

→L.Val.L.Orn.L.Leu.X.L.Pro→
←L.Tyr.L.Gln.L.Asn.Z.Y←

(225)

The synthesis of the individual peptide chains of antibiotics can be accomplished in the ways discussed in preceding chapters. Cyclic peptides can be synthesized by a head to tail condensation of these chains using any of the available activating methods. However, special regard must be given to two points; the first concerns the position of ring-closure and the second, the conditions which prevail during the cyclization reaction. Ideally, the linear peptide should be chosen so that its *C*-terminal residue is in no danger of racemizing when it is activated for the ring-closure step. With gramicidin S, therefore, the bond between proline and valine is most suitable from this point of view. The actual cyclization should be carried out in dilute solution to minimize the formation of dimers, trimers etc. This is usually achieved by utilizing an acid-labile protecting group for the terminal amino group so that a protonated species is produced when the protecting group is removed. Since no further reaction can occur whilst the amino group is present as a salt, it is an easy matter to ensure that cyclization does not occur until the solution has been suitably diluted. Gramicidin S has been synthesized by the route summarized in Scheme 6.1.

It is found that the pentapeptide *p*-nitrophenyl ester also gives gramicidin S when cyclized in the way indicated. Similar doubling reactions have been observed with other peptides during ring closures, but they do not always occur, as here, to the complete exclusion of straight-forward cyclization. With small peptides, for example L-prolyl-L-prolylglycine, dimerization suggests that there is prohibitive strain or hindrance in the cyclization transition state. With larger peptides, dimerization presumably occurs because association in the transition state is particularly favoured. Association of this type is highly specific. For example, cyclosemigramicidin S can be obtained by cyclizing the N^{δ}-benzyloxycarbonylornithine pentapeptide *p*-nitrophenyl ester in pyridine solution in the manner shown in Scheme 6.1.

The use of *N*,*N*′-dicyclohexylcarbodiimide at the cyclization stage is exemplified by the synthesis (Scheme 6.2) of polymyxin B_1 (**226**; MOA = (+)-6-methyloctanoic acid; Dab = α,γ-diaminobutanoic acid). This antibiotic is a member of a large family of α,γ-diaminobutanoic acid-containing compounds, isolated from *Bacillus polymyxa*.

L.Thr.L.Dab.L.Dab.L.Leu

MOA.L.Dab.L.Thr.L.Dab.L.Dab.L.Dab.D.Phe

(**226**)

The aminoacyl insertion reaction (p. 39 *et seq*) has been harnessed for the synthesis of model cyclic peptides. Thus, hydrogenolysis of *N*-benzyloxycarbonyl-β-alanyllactams (**227**; $n = 1$ or 3) gives the resulting cyclopeptides (**228**). Two factors are responsible for the reactivity of diacylamides. One is the electron-withdrawing effect of the second carbonyl group which disturbs the resonance of the simple amide; the other is the disturbance of the planarity of the amide bond (see p. 154).

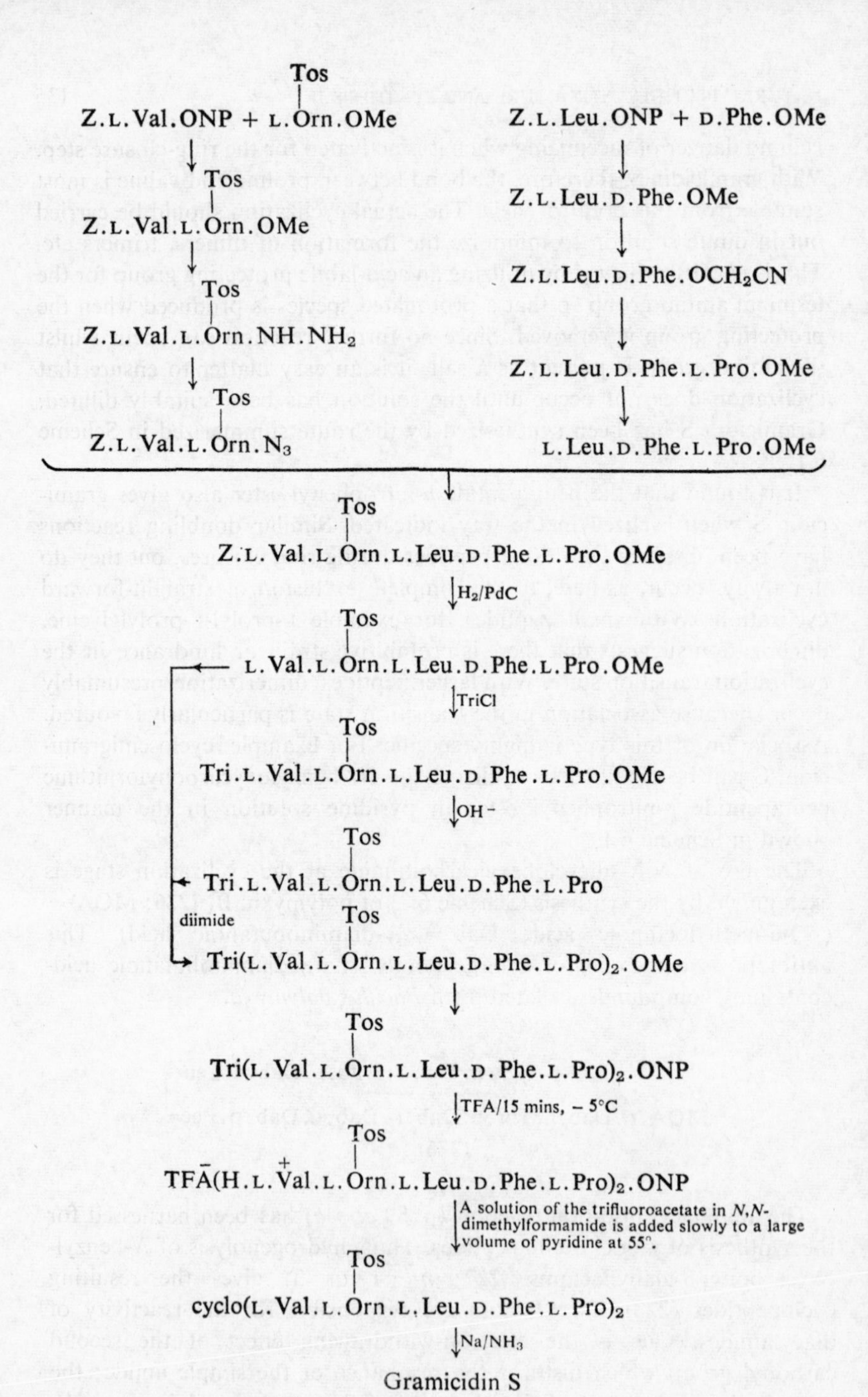

Scheme 6.1. Synthesis of gramicidin S (Z = $PhCH_2OCO$; Tos = $CH_3C_6H_4.SO_2(p)$; TFA = CF_3CO_2H; ONP = $OC_6H_4NO_2(p)$; Tri = Ph_3C)

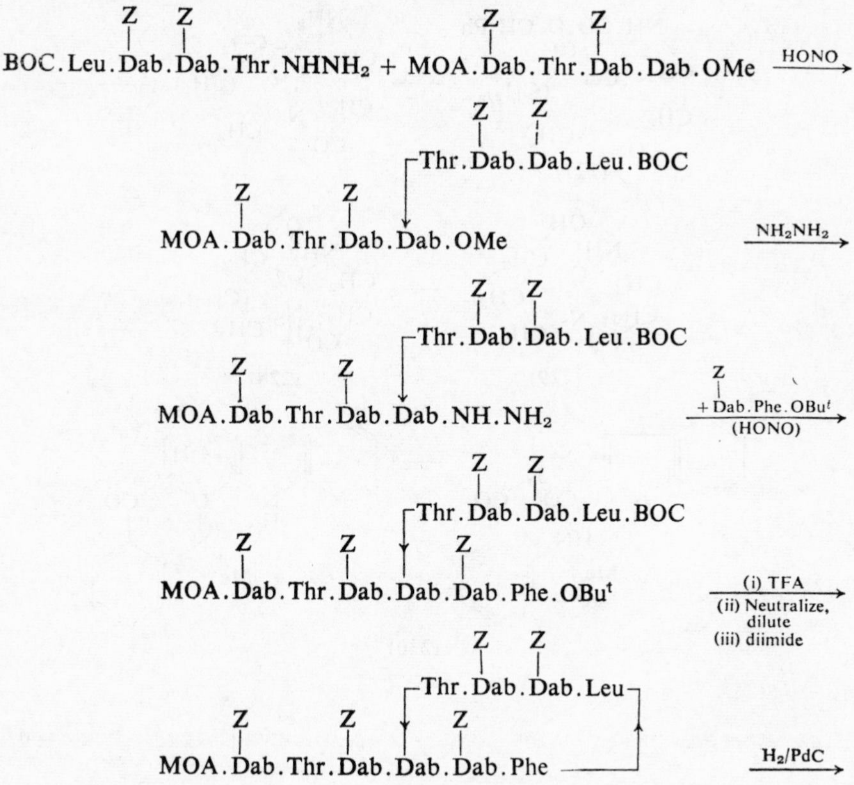

Z Z Z Z

BOC.Leu.Dab.Dab.Thr.$NHNH_2$ + MOA.Dab.Thr.Dab.Dab.OMe $\xrightarrow{HONO}$

Z Z

Thr.Dab.Dab.Leu.BOC

Z Z

MOA.Dab.Thr.Dab.Dab.OMe $\xrightarrow{NH_2NH_2}$

Z Z

Thr.Dab.Dab.Leu.BOC

Z Z

MOA.Dab.Thr.Dab.Dab.NH.NH_2 $\xrightarrow[(HONO)]{+\ \text{Dab.Phe.OBu}^t\ (Z)}$

Z Z

Thr.Dab.Dab.Leu.BOC

Z Z Z

MOA.Dab.Thr.Dab.Dab.Dab.Phe.OBu^t $\xrightarrow{}$ (i) TFA (ii) Neutralize, dilute (iii) diimide

Z Z

Thr.Dab.Dab.Leu

Z Z Z

MOA.Dab.Thr.Dab.Dab.Dab.Phe $\xrightarrow{H_2/PdC}$

Polymyxin B_1

Scheme 6.2. Synthesis of Polymyxin B_1 (BOC = $CMe_3.O.CO$; Z = $PhCH_2OCO$; MOA = (+)-6-methyloctanoic acid); Dab = α,γ-diaminobutanoic acid)

The intermediate in the rearrangement is called a *cyclol* (**229**). Early theories of peptide structure postulated an appreciable contribution from forms of this type. In the event, such structures are usually very reactive and are generally only observed as transient intermediates, but in one or two cases cyclols of this type are thought to exist, as, for example, in the alkaloid rhetsinine (**230**). The possibility of their formation must always be taken into account when considering the reactivity of peptides, particularly of cyclopeptides where transannular reactions can be sterically and energetically favoured.

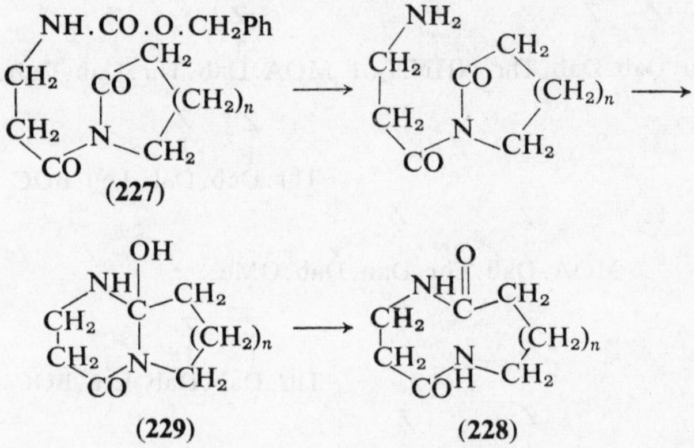

(227)

(229) (228)

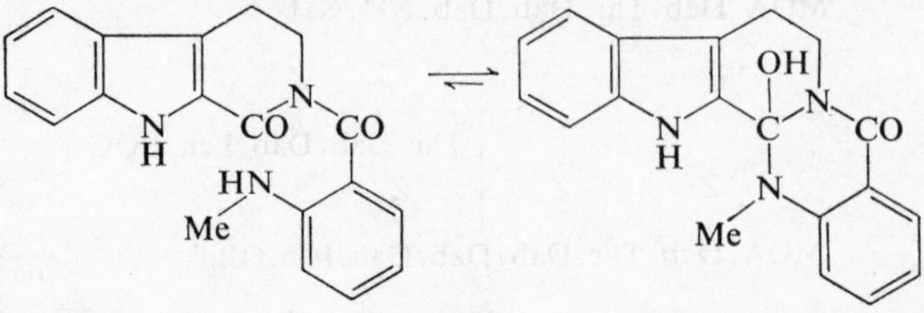

(230)

*Exercise 6.8**

Etamycin [$C_{44}H_{62}N_8O_{10}$; $M = 890$ by potentiometric titration with sodium hydroxide; ultraviolet: $\lambda_{max} = 303$ nm in 0·1M HCl; $\lambda_{max} = 350$ nm in 0·1M NaOH] is soluble in benzene and carbon tetrachloride. It can be hydrolysed to give equimolar amounts of L-threonine, L-alanine, D-leucine, sarcosine (*N*-methylglycine), L-α-phenylsarcosine (Exercise 6.4*), L-β,*N*-dimethylleucine (Exercise 6.5*), D-γ-allo-hydroxyproline (a-Hypro) and Compound A.

Compound A [$C_6H_5O_3N$; ultraviolet: $\lambda_{max} = 301·5$ nm in 0·1M HCl] gives a brownish-red colour with ferric chloride. When heated strongly it decomposes to give 3-hydroxypyridine. Etamycin [infrared: $\nu_{max} = 1745$ cm^{-1}] can be hydrolysed with aqueous sodium hydroxide to give a disodium salt (infrared: $\nu_{max} = 1730$ cm^{-1}). Both threonine and hydroxyproline residues in this material are destroyed by treatment with chromic acid whereas the threonine remains unchanged when the starting material is treated in this way. Catalytic hydrogenation of etamycin followed by treatment with alkali gives a 3-hydroxypipecolic acid (Hypic; **231**) derivative. Hydrazinolysis of this compound yields α-phenylsarcosine and Edman degradation indicates the following sequence of residues:

Hypic.Thr.Leu.a-Hypro.Sar.Me_2Leu.Ala.

Elucidate the structure of etamycin.

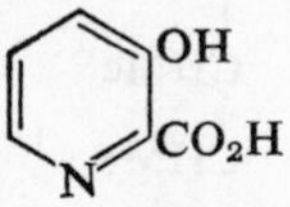

OH
CO_2H
N

(231)

*Exercise 6.9**

Staphylomycin S [M = 823, infrared: ν_{max} = 1742 cm^{-1}] can be hydrolysed to give 3-hydroxypicolinic acid, L-threonine, D-α-aminobutanoic acid, L-proline, L-*N*-methylphenylalanine, L-4-oxopipecolic acid and L-α-phenylglycine. The mass spectrum of staphylomycin S possesses diagnostic peaks at m/e 778, 673, 548, 387, 290, 205, 122. Deduce the structure of staphylomycin S.

Heterodetic cyclic peptides

Etamycin, an antibiotic extracted from *Streptomyces* species, is another type of cyclic peptide. In this molecule (**232**), the C-terminal α-phenylsarcosine residue is linked ester-wise to the hydroxyl group of the *N*-terminal threonine residue and the terminal amino group is acylated by 3-hydroxypicolinic acid. Structures of this type, in which other than peptide bonds are present in the ring, are called heterodetic cyclic peptides. Etamycin is properly styled a heteromeric heterodetic cyclic peptide. By the same reasoning, oxytocin is a homomeric heterodetic cyclic peptide.

Streptomyces species have proved the most prolific source of antibiotics and many of their products possess a macrocyclic lactone type of structure. Staphylomycin S (**233**), for example, is very similar to etamycin.

OH

OH

(L) (D) D(allo)

N CO.NH.CH.CO.Leu.N—CH.CO

CO.O.CH.Me

Ph.CH.N.Ala.CO.CH.NMe.Sar

(L) Me (L)

CHMe

$CHMe_2$

(232)

OH Et

(L) (D)

N CO.NH.CH.CO.NH.CH.CO.N—CH.CO

CO.O.CH.Me

Ph.CH.CO—CH—N—CO.CH.N.Me

(L)

CH_2.Ph

O

(233)

The structure of this compound was first determined by chemical degradation, but it does, in fact, provide an excellent illustration of the potential value of mass spectrometry (Exercise 6.9*) in this field. In general, these lactones tend to be somewhat more volatile than homodetic peptides and mass spectra can be obtained from them more readily. Primary fragmentation occurs at the lactone group (**234** → **235**) followed by stepwise elimination of the amino acid residues (**235** → **236** → **237** etc.) in the usual way (p. 34).

Thus, the peaks in the mass spectrum of staphylomycin S may be assigned unambiguously (**238**).

Other antibiotics, for example, the ostreogrycins are structurally very similar to etamycin, but some are much different. Echinomycin (**239**), with its bilactone structure bridged by a dithiolane ring, is considerably more complex. In this case, the acylating group is quinoxaline-2-carboxylic acid.

Probably the best known peptide antibiotics are the actinomycins, which are also obtained from various streptomyces species. These compounds possess a phenoxazine chromophore to which two peptide lactones are attached. See, for example actinomycin C_3 (**240**).

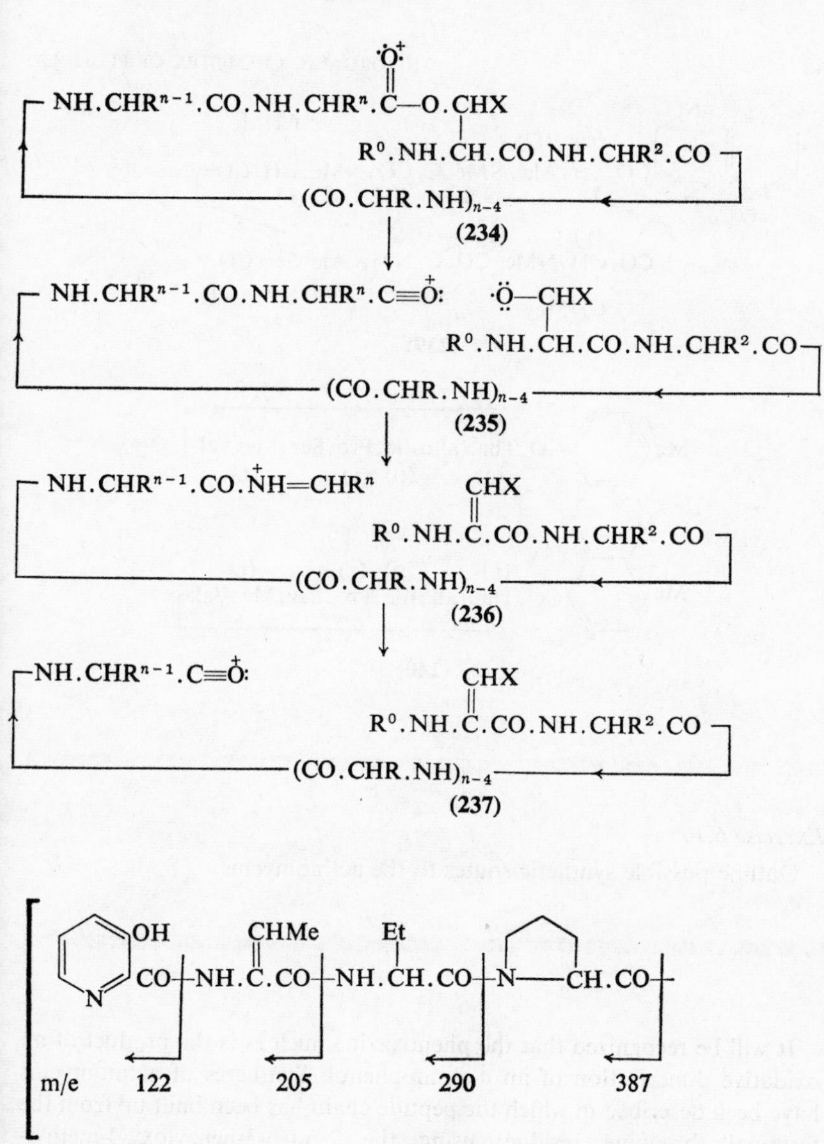

O

CH2Ph

N.CH.CO—N—CH.CO—NH=CHPh

Me

548 673 778

(238)

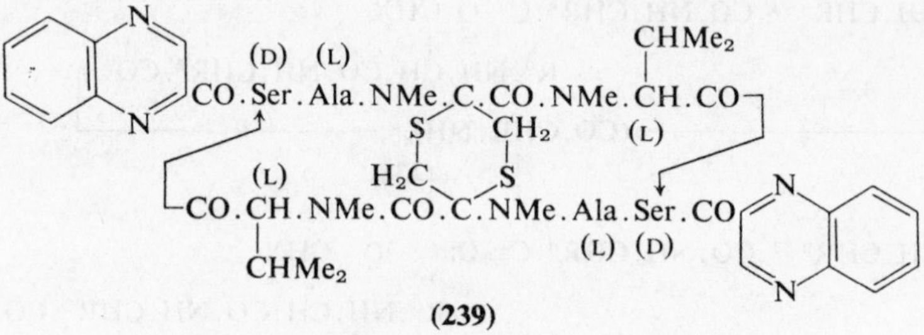

N
CHMe$_2$
(D) (L)
CO.Ser.Ala.NMe.C.CO.NMe.CH.CO
N S CH$_2$ (L)
(L) H$_2$C S N
CO.CH.NMe.CO.C.NMe.Ala.Ser.CO
(L) (D)
CHMe$_2$ N

(239)

Me CO.Thr.(allo)Ile.Pro.Sar.Me.Val
(L) (D) (L) (L)
O N
(L) (D) (L) (L)
Me CO.Thr.(allo)Ile.Pro.Sar.Me.Val
O

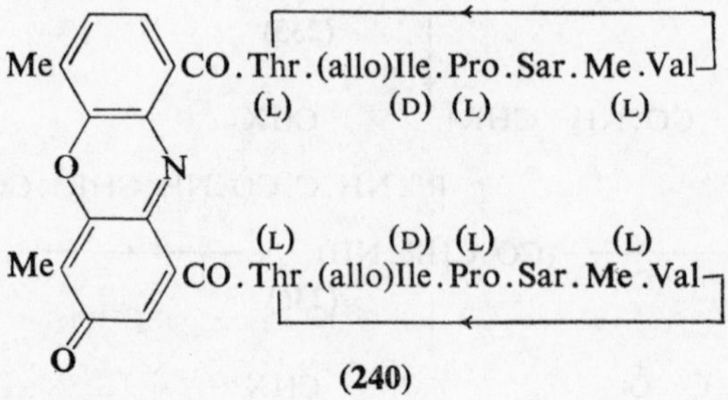

(240)

*Exercise 6.10**

Outline possible synthetic routes to the actinomycins.

It will be recognized that the phenoxazine nucleus is the product of an oxidative dimerization of an *o*-aminophenol. Syntheses of actinomycins have been described in which the peptide chain has been built up from the terminal threonine residue, using the 2-nitro-3-benzyloxy-4-methylbenzoyl group to protect the terminal amino group (**241** → **242**). The nitro group has then been reduced, with concomitant cleavage of the benzyl ether, and the resulting 2-amino-3-hydroxy-4-methylbenzoyl peptides (**243**) have been oxidized with ferricyanide to the phenoxazine derivatives (**244**). Final ring closure has involved either peptide bond formation, for example with diimide, or lactonization, for example with acetyl chloride in pyridine (mixed anhydride). Syntheses have also been described in which the oxidation of the *o*-aminophenol was the final step.

Me CO.Thr ⟶ Me CO.Thr.(allo)Ile.etc. ⟶
$PhCH_2O$ NO_2 $PhCH_2O$ NO_2
(241) **(242)**

Me CO.Thr.(allo)Ile.etc ⟶ Me CO.Thr.(allo)Ile.etc
HO NH_2 O N
(243)
Me CO.Thr.(allo)Ile.etc
O NH_2
(244)

---

*Exercise 6.11**

Enniatin A [$C_{36}H_{63}O_9N_3$; infrared: ν_{max} = 1736, 1661, 1167 cm^{-1}], a fat-soluble antibiotic isolated from *Fusarium* species, gave by total acidic hydrolysis D-α-hydroxyisovaleric acid and compound I, $C_7H_{15}NO_2$, in equimolar amounts. *N*-Methylamine and (+)-2-methylbutanal were obtained by oxidation of compound I. Acidification of a neutral solution of compound I made its optical rotation become more positive.

Partial alkaline hydrolysis of enniatin A gave compound II, $C_{12}H_{23}NO_4$, which did not give a colour with ninhydrin reagent. Distillation of II yielded compound III, $C_{12}H_{21}NO_3$, which could be hydrolysed under acidic conditions to D-α-hydroxyisovaleric acid and compound I. Explain these reactions and propose structures for compounds I, II and III and for enniatin A. Suggest methods by which compound I and enniatin A could be synthesized to confirm the proposed structures.

---

*Exercise 6.12**

Sporidesmolide I [$C_{33}H_{58}N_4O_8$; infrared: ν_{max} = 3350, 2968, 2930, 1753, 1679, 1646, 1529 cm^{-1}], a toxic factor isolated from the fungus,

Pithomyces chartarum, yields by total acidic hydrolysis L-α-hydroxyisovaleric acid, D-valine, L-valine, *N*-methyl-L-leucine and D-leucine in molar ratios 2:1:1:1:1. Partial alkaline hydrolysis, in which two equivalents of alkali are consumed, gives two ninhydrin-negative compounds A and B. Compound A, $C_{16}H_{30}N_2O_5$, by total hydrolysis, goes to L-α-hydroxyisovaleric acid, D-valine and D-leucine. Treatment with hot pyridine and acetic anhydride prior to the hydrolysis, a procedure known to convert amino acids and *C*-terminal amino acids to methyl ketones (Dakin–West reaction), shows that leucine occupies the *C*-terminal position.

Compound B, $C_{17}H_{32}N_2O_5$, gives by total hydrolysis, L-α-hydroxyisovaleric acid, D-valine and *N*-methyl-L-leucine. In this instance, the latter amino acid was missing in hydrolysates of the Dakin–West reaction product.

Deduce the structure of sporidesmolide I.

---

*Exercise 6.13**

Pithomycolide [$C_{30}H_{36}N_2O_8$; $M = 535 \pm 20$ by X-ray; infrared; $\nu_{max} = 1739, 1718, 1669, 1642, 1534\ cm^{-1}$], also isolated from *Pithomyces chartarum*, gives, by total acidic hydrolysis, L-α-hydroxyisovaleric acid, L-alanine, *N*-methyl-L-alanine and cinnamic acid in molar ratios 1:1:1:2. D-(*S*)-β-Hydroxy-β-phenylpropanoic acid can be detected in partial alkaline hydrolysates. Up to 1·4 moles of aromatic acids are produced by mild alkaline hydrolysis. Borohydride reduction of pithomycolide followed by total hydrolysis, leads to the formation of 2-aminopropanol and *N*-methylalanine. Account for these observations and suggest a structure for pithomycolide.

---

*Exercise 6.14**

Angolide, a crystalline compound isolated from a species of *Pithomyces*, has been accorded structure (**245**) on the basis of chemical degradative studies. The mass spectrum of angolide shows, *inter alia*, the following peaks: m/e 426, 382, 367, 297, 282, 197, 182, 69. Are these in agreement with the proposed structure?

```
   Me2CH                         CHMeEt
        \                       /
    (L) CH—CO—NH—CH (L)
         |                      |
         O                     CO
         |                      |
        CO                      O
         |                      |
(D-allo) CH—NH—CO—CH (L)
        /                       \
  MeEtCH                         CHMe2
                (245)
```

Depsipeptides

Microorganisms have now yielded many compounds which can be hydrolysed to α-amino acids and hydroxyacids and, from the manner in which these compounds can be hydrolysed and saponified, it is apparent that they contain both peptide and ester bonds. Such compounds are called depsipeptides. They are generally neutral, and possibly always cyclic, i.e. cyclodepsipeptides. Infrared spectroscopy provides valuable information in the investigation of depsipeptides because of the ease with which it can be used to detect ester carbonyl (1755–1715 cm^{-1}), amide carbonyl (1680–1635 cm^{-1}) and amide NH (1575–1500 cm^{-1}) groups. The chemical reactions which these compounds undergo can be summarized as (a) saponification of the ester bonds (**246** → **247**); and (b) total hydrolysis in which both ester and amide bonds are cleaved (**246** and **247** → **248** and **249**). The hydroxyacylamino acid derivatives (**247**) cyclize readily on

cyclo-(—O—CHR^1.CO.NH.CHR^2.CO—$)_n$ **(246)**

↓ OH^-

HO.CHR^1.CO.NH.CHR^2.CO_2^- **(247)**

(246) —H^+/H_2O→ (248) + (249)

(247) —H^+/H_2O→ HO.CHR^1.CO_2H **(248)** + $\overset{+}{N}H_3$.CHR^2.CO_2H **(249)**

(247) ↓ heat

```
     CHR1
   /     \
 OC       O
  |       |
 HN       CO
   \     /
     CHR2
```
(250)

(250) —H^+/H_2O→ (248) + (249)

heating in a manner reminiscent of 2,5-dioxopiperazine formation (p. 14) and the resulting 2,5-dioxomorpholines (**250**) can be hydrolysed to their constituent hydroxy and amino acids.

Occasionally, side reactions occur, like the elimination of a β-hydroxyl group from β-hydroxy-β-phenylpropanoic acid (note carbonium ion stabilization) (Exercise 6.13*), but the sequence of the residues can usually be deduced by a simple investigation of the hydrolysis products. Thus, the structures of enniatin A (**251**), sporidesmolide I (**252**) and pithomycolide (**253**) follow from evidence of this kind (Exercises 6.11*–6.13*).

Unfortunately, as with simple peptides there are often serious practical problems in studying depsipeptides. For example, a case has been recorded of two synthetic diastereoisomers which were indistinguishable by melting point, mixed melting point, infrared spectroscopy, X-ray diffraction pattern (powder), mass spectrometry and chromatography. The purification of

cyclo-(—O—CH($CHMe_2$)—CO—NMe—CH(CHMeEt)—CO—)$_3$

(**251**)

cyclo-[O.CH($CHMe_2$).CO.Val.Leu.CH(CHMe$_2$)(L).O.CO.Val(L).NMe.CH(CH_2CHMe_2)(L).CO] — with Val (D), Leu (D)

(**252**)

cyclo-[NMe.CH(Me).CO.Ala.O.CH(Ph).CH$_2$.CO.O.CH(Ph).CH$_2$.CO.Val] — (L), (L), (D), (L)

(**253**)

such compounds is obviously difficult. In addition, solution measurements with cyclodepsipeptides sometimes suggest erroneous molecular weights. Electrothermal, X-ray crystallographic and mass spectrometric measurements are invaluable from this point of view.

In the mass spectrometer cyclodepsipeptides undergo ring opening at an ester bond and, thereafter, stepwise fragmentation like peptides (**254** → **255** → **256** → **257**; X = 0, NH or NMe; Y = H; Z = Me, Et, $CHMe_2$). However, the degradation is often complex due to the number of places at which the initial ring-opening can occur. Usually, two further types of fragmentation are also observed. These involve either two amide or two ester groups respectively and result in the formation of 2,5-dioxomorpholine (**258**) and acylaminoketene (**259**) fragments.

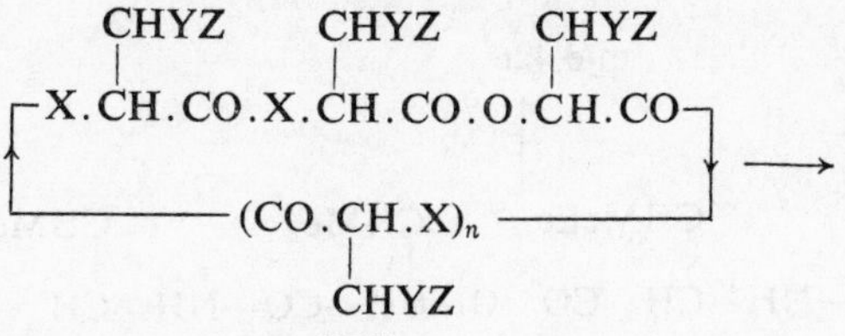

```
  CHYZ        CHYZ        CHYZ
   |           |           |
┌─X.CH.CO.X.CH.CO.O.CH.CO─┐
↑                          ↓   ⟶
└──────── (CO.CH.X)n ──────┘
              |
             CHYZ
            (254)

  CHYZ          CHYZ CY(or Z)
   |        +    |    ||
┌─X.CH.CO.X═CH     CH.CO─┐
↑                         ↓  ⟶
└──────── (CO.CH.X)n ─────┘
              |
             CHYZ
            (255)

  CHYZ              CY(or Z)
   |        +        ||
┌─ X.CH.C≡O:         CH.CO ─┐                 CY
↑                            ↓  - - - ->      ||   +
└──────── (CO.CH.X)n ────────┘               CH.C≡O:
               |                             (257)
              CHYZ
             (256)

YZHC                                         ]+
    \   X                [CH.CO.X.C═C═O
     CH   CO              ||       |
     |    |               CYZ      CHYZ      ]
     OC   CH
        X    \
              CHYZ
   (258)                   (259)
```

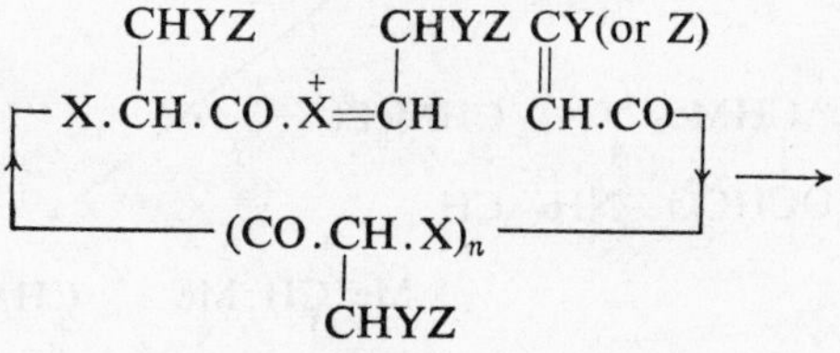

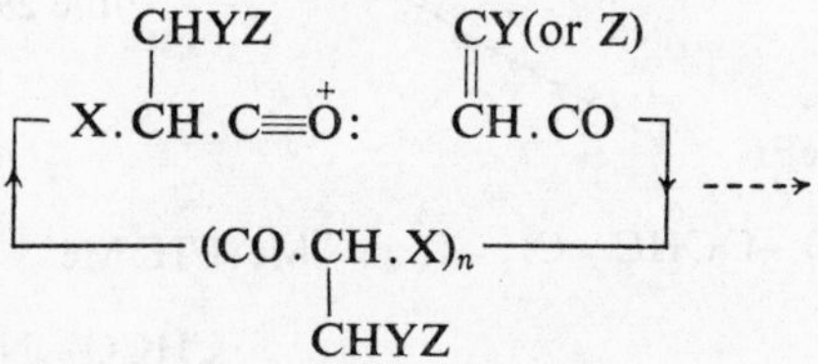

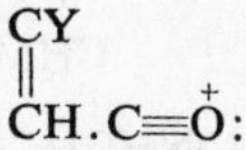

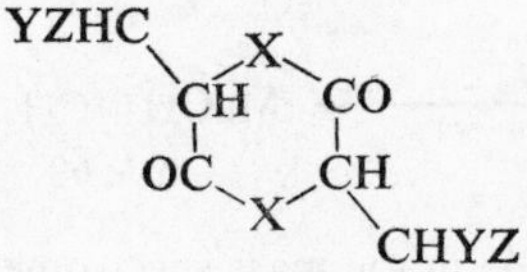

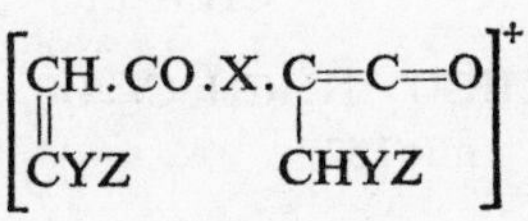

Complete sequences can often be deduced on the basis of mass spectra. The main fragmentation pattern of angolide (Exercise 6.14*) is illustrated (Scheme 6.3).

The usual protecting groups have been employed in the synthesis of depsipeptides. In one case, the amino group of *N*-methyl-L-isoleucine was protected as its *N*-nitroso derivative, from which the free amine was recovered by treatment with hydrogen chloride in benzene. Generally, ester bonds are formed first because cyclization with peptide bond formation has often proved more efficient than lactonization. Since hydroxyl groups are much weaker nucleophiles than amino groups, stronger activation of the carboxylic group is necessary during ester

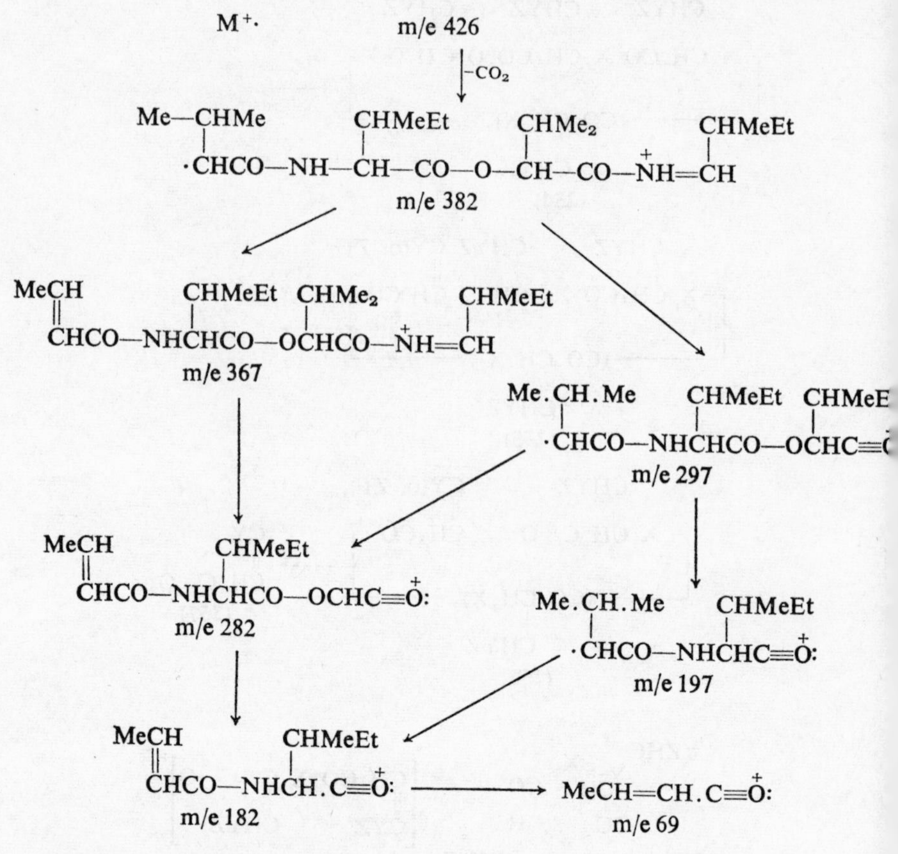

Scheme 6.3. Fragmentation of angolide in the mass spectrometer

formation. When *N*-methylamino acid residues are present, strong activation of the carboxylic group is also necessary for the formation of the amide bond. Perhaps because of steric hindrance, *N*-methylamino acids do not make good nucleophiles. Couplings are usually carried out with acid chlorides, formed with phosphorus pentachloride or with thionyl chloride; or with mixed anhydrides, for example formed between the carboxylic acid and benzene sulphonyl chloride in the presence of pyridine. It is fortunate that oxazolone formation cannot occur with these compounds. Cyclization is brought about at high dilution, for example, by neutralization of the protonated open chain acid chloride as in the synthesis of enniatin A (Scheme 6.4). Cyclodimerizations of depsipeptides have been observed.

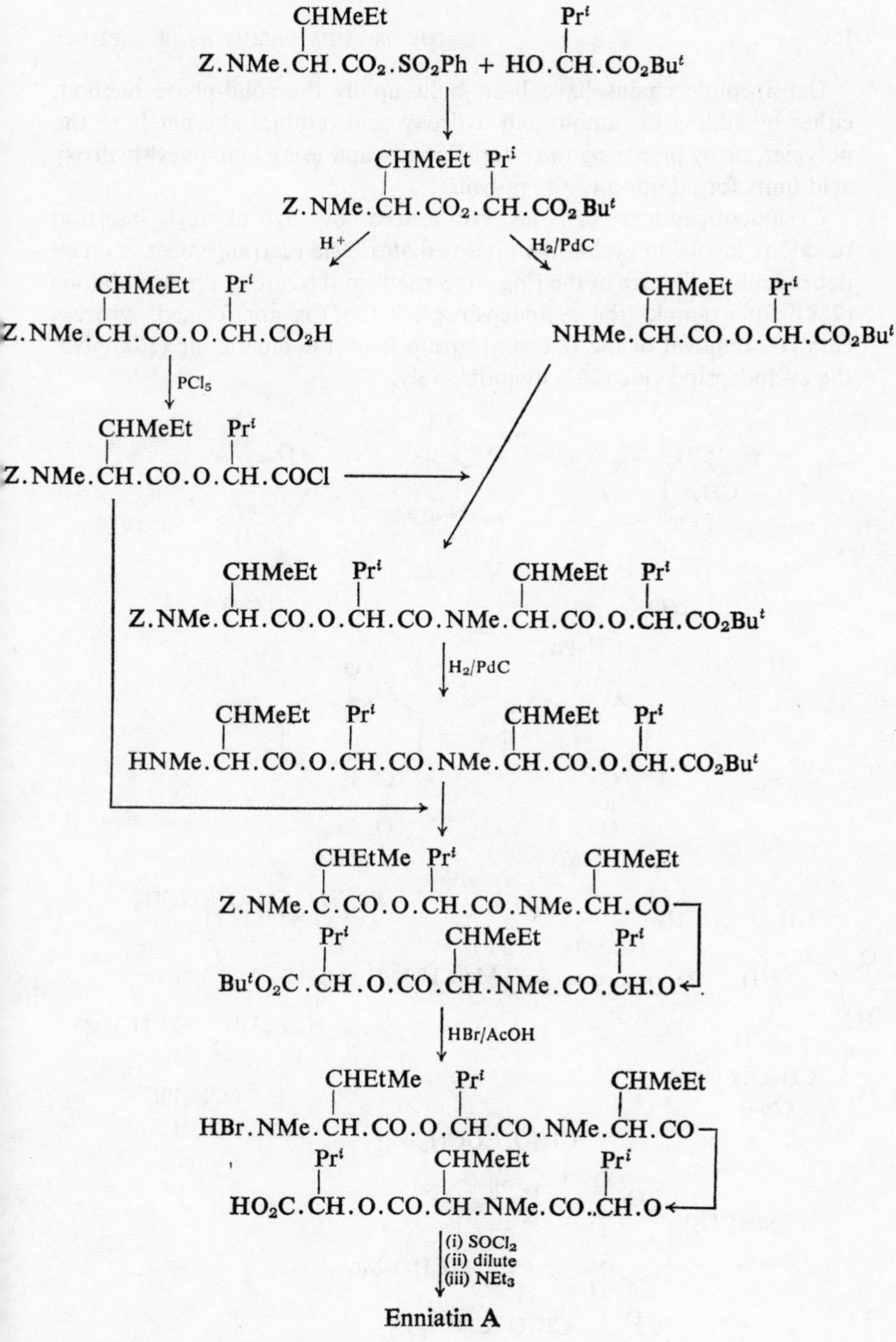

$Z.NMe.CH(CHMeEt).CO_2.SO_2Ph + HO.CH(Pr^i).CO_2Bu^t$

↓

$Z.NMe.CH(CHMeEt).CO_2.CH(Pr^i).CO_2Bu^t$

H^+ ↙ ↘ H_2/PdC

$Z.NMe.CH(CHMeEt).CO.O.CH(Pr^i).CO_2H$ $\qquad$ $NHMe.CH(CHMeEt).CO.O.CH(Pr^i).CO_2Bu^t$

↓ PCl_5

$Z.NMe.CH(CHMeEt).CO.O.CH(Pr^i).COCl$ →

↓

$Z.NMe.CH(CHMeEt).CO.O.CH(Pr^i).CO.NMe.CH(CHMeEt).CO.O.CH(Pr^i).CO_2Bu^t$

↓ H_2/PdC

$HNMe.CH(CHMeEt).CO.O.CH(Pr^i).CO.NMe.CH(CHMeEt).CO.O.CH(Pr^i).CO_2Bu^t$

↓

$Z.NMe.C(CHEtMe).CO.O.CH(Pr^i).CO.NMe.CH(CHMeEt).CO$ —┐

$Bu^tO_2C.CH(Pr^i).O.CO.CH(CHMeEt).NMe.CO.CH(Pr^i).O$ ←┘

↓ HBr/AcOH

$HBr.NMe.CH(CHEtMe).CO.O.CH(Pr^i).CO.NMe.CH(CHMeEt).CO$ —┐

$HO_2C.CH(Pr^i).O.CO.CH(CHMeEt).NMe.CO.CH(Pr^i).O$ ←┘

↓ (i) $SOCl_2$ (ii) dilute (iii) NEt_3

Enniatin A

Scheme 6.4. Synthesis of Enniatin A ($Z = PhCH_2OCO$)

Depsipeptide chains have been built up by the solid-phase method, either by adding the amino and hydroxy acid residues alternately to the polymer, or by preparing the esters initially and using aminoacyl-hydroxy acid units for addition to the polymer.

Cyclodepsipeptides can be synthesized by hydroxyacyl insertion reactions involving cyclol-like intermediates. The rearrangement is much dependent on the size of the ring concerned; in the case of the pyrrolidone (**260**), for example, the cyclodepsipeptide (**261**) is not formed, whereas catalytic removal of the *O*-benzyl group from the piperidone (**262**) gives the cyclodepsipeptide (**263**) quantitatively.

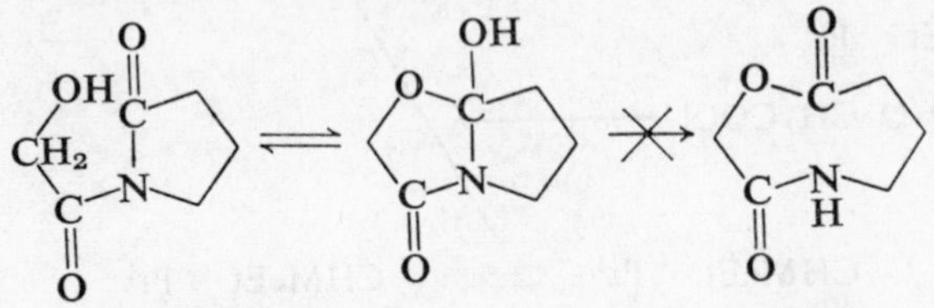

(260) **(261)**

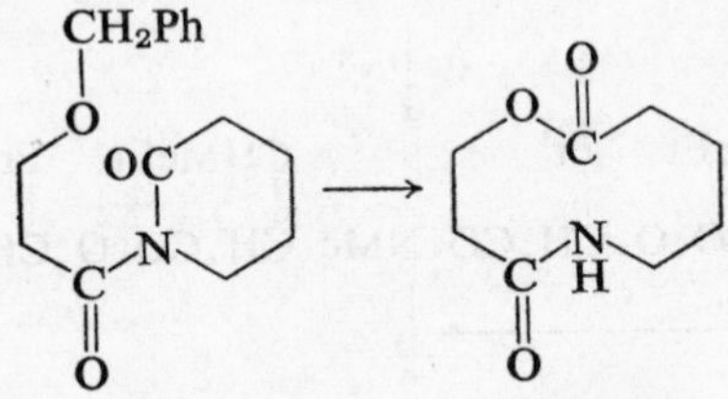

(262) **(263)**

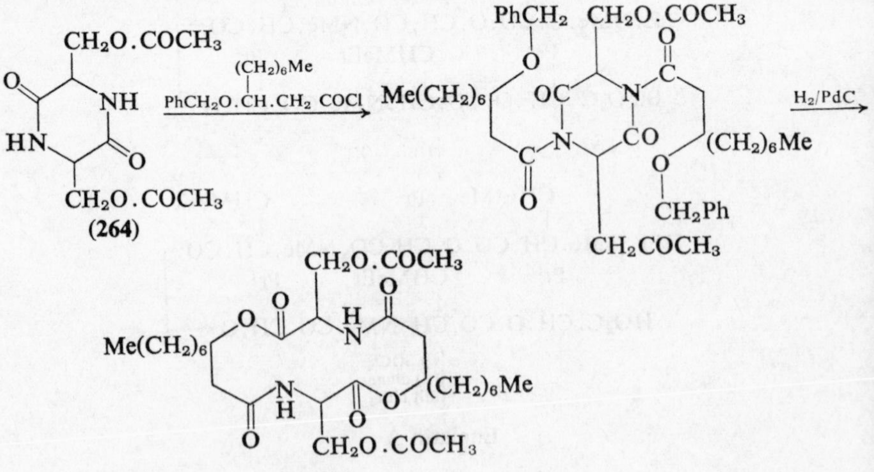

$CH_2O.COCH_3$

$PhCH_2O.\overset{(CH_2)_6Me}{CH}.CH_2COCl$

H_2/PdC

(264)

(265)

The *O,O'*-diacetyl derivative of serratamolide, a metabolite of *Serratia marcescens*, has been synthesized in this way from the 2,5-dioxopiperazine of L-serine (**264** → **265**).

Exercise 6.15

The tetrabenzyl ether (**266**) does not give a cyclodepsipeptide on hydrogenolysis. Instead, a high yield of the side chain substituted *O,O'*-bis(β-hydroxypropanoyl)-dioxopiperazine derivative is obtained. Trace the course of the reactions involved.

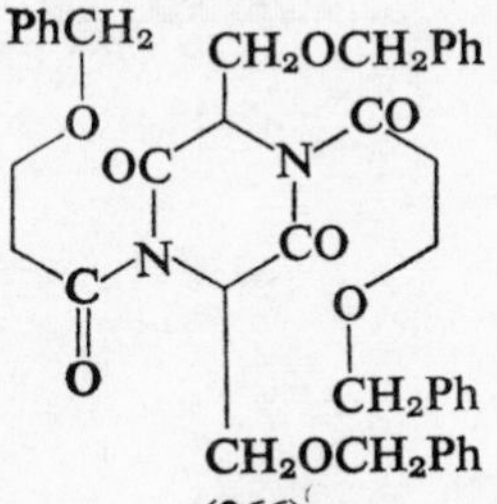

(266)

Oxacyclols (e.g. **267**) have been obtained crystalline and characterized spectroscopically. Many of them tend to rearrange to the corresponding hydroxyacyl or depsipeptide derivatives, but some are relatively stable. A cyclol structure is undoubtedly present in the ergot peptide (**268**) and

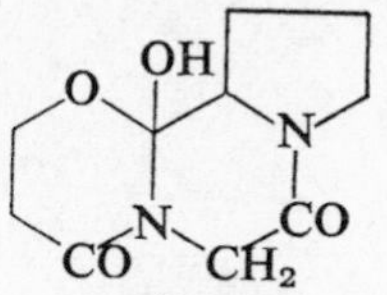

(267)

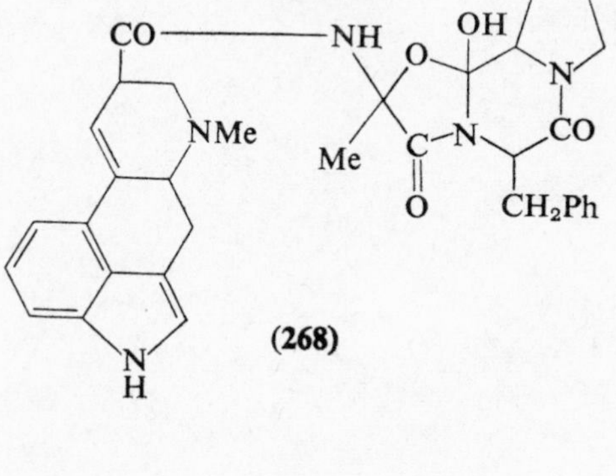

(268)

cyclol structures have been postulated to account for the unusual reactivity of other peptides.

Exercise 6.16

Suggest a method for the synthesis of the cyclol (**267**).

CHAPTER SEVEN

Structure and Biological Function

The importance of conformation

One of the most striking things about peptides is the tremendous range of biological functions which they perform. Naturally, this diversity of function is paralleled by an equally broad range of properties. Thus, fibrous proteins possess quite appreciable mechanical strength, are insoluble in water and, chemically, are relatively inert; whereas globular proteins and smaller natural peptides are soluble in water and have characteristic reactivity.

To a limited extent, differences in function are reflected in differences in molecular weight. Fibrous proteins generally possess very high, somewhat indefinite, molecular weights and are sometimes polymer-like. Globular proteins have much smaller molecular weights and exist as discrete chemical entities, often obtainable in crystalline form. However, molecular weight differences are not the whole story. Nor is it possible to account for the properties of peptides solely in terms of primary structure which conveys nothing of the three dimensional arrangement of the amino acid residues in the molecule. Due to the convolutions of the peptide chains, amino acid residues which seem to be widely separated in the primary structure may in reality be close together and able to act in concert. Other residues which seem to be buried in the primary structure, may in fact be at the surface of the peptide molecule and thus influence in a profound way its interaction with the surrounding medium.

Two further levels of structural complexity are defined. The secondary structure of a peptide is the way in which the individual peptide chains are coiled; its tertiary structure, the way in which the coiled chains are folded. Both must be taken into account if a meaningful correlation is to be made between structure and function.

The forces which determine conformation

*Exercise 7.1**

It is instructive at this stage to build a simple tripeptide from structural models to see the apparent degrees of flexibility involved. Remember that with longer chains, the probability of intrachain interactions will become greater.

*Exercise 7.2** †

At room temperature, all of the methyl protons in *N*-methylacetamide, *N*-methylpropanamide and *N*-methyl-2-methylpropanamide are accounted for by a multiplet which centres on 7·3τ, whereas the methyl protons of *N*-methylformamide appear as two groups of multiplets at 7·3τ and 7·13τ respectively in the approximate ratio 92:8. Suggest an explanation for these differences.

† See appendix.

*Exercise 7.3**

List the factors and forces which might influence the conformation of a polypeptide chain.

Because of mesomerism (**269**) the amide bonds in the peptide chain tend to be planar. That is to say, the α-carbon atoms attached to the CO–NH skeleton of a given amide tend to lie in the same plane as the C, O, N and H atoms.

$$R.\overset{O^-}{\overset{|}{C}}=\overset{+}{N}H.R' \longleftrightarrow R.\overset{O}{\overset{\|}{C}}-NH.R' \qquad \left[R.\overset{O}{\overset{\vdots}{C}}\cdots NH.R'\right]$$

(269)

Whilst both *cis* (**270**) and *trans* (**271**) forms of the amide can be visualized, the *trans* form is favoured thermodynamically by more than 2 kcal/mole.

O, H, C^α, C′=N, C^α (cis) C^α, H, C′=N, O, C^α (trans)

(270)   **(271)**

Deviation from planarity in the amide bond, defined by the angle ω which the plane C^α–C′–N makes with the plane C′–N–C^α, is associated with a destabilization energy of the form $K\omega^2$, where K is of the order of 15–30 kcal/mole/degree^2. Consequently, in most peptides, the amide bonds are of the planar *trans* type and exceptions only occur in unusual situations, for example, in small cyclic peptides and in dioxopiperazines.

When considering the conformation of a peptide, it is convenient to regard the amide (**272**), rather than the amino acid residue (NH.CHR.CO), as the structural unit. The full stereochemistry of the peptide backbone can then be defined by the relative orientations of the N–C^α and C^α–C′ bonds in consecutive units. These orientations are usually defined in terms of the clockwise angular rotations (ϕ and ψ) of the N–C^α and C^α–C′ bonds from the fully extended planar position (Figure 7.1).

C^α — 1·53 Å — C′; C′ — 1·32 Å — N; C′ — 1·24 Å — O; N — 1·00 Å — H; N — 1·47 Å — C^α; angles: 121° (at C′), 125° (at C′), 123° (at N), 114° (at N), 123° (at N)

(272)

Since the lengths of the N–C^α and C^α–C′ bonds approximate to single bond distances, it is to be expected that free rotation will be possible along these bonds. Similarly, the C^α–H bonds and most of the bonds involved in the various side chains of the amino acid residues, with the exception of bonds which form parts of aromatic moieties, are single bonds which

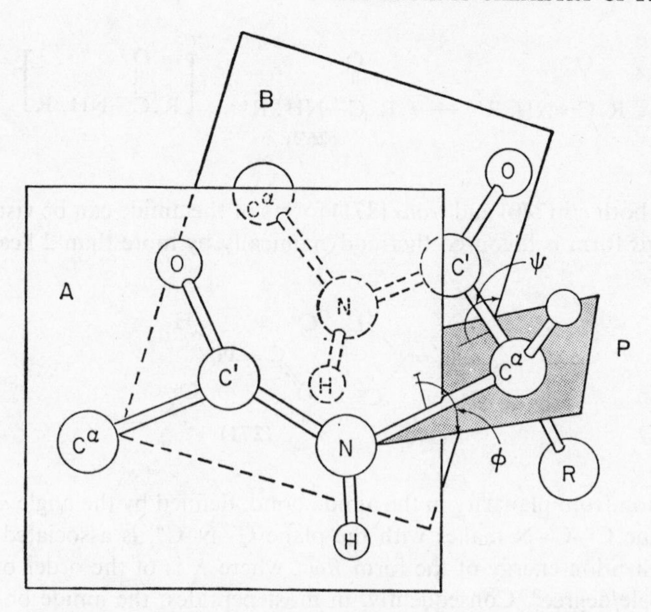

Figure 7.1. Designation of conformation. The atoms C′, C^{α} and N all lie in the shaded plane P; planes A and B contain the two adjacent amide units. Clockwise rotation ϕ of plane A brings it to plane P and clockwise rotation ψ of plane P brings it to plane B (From G. N. Ramachandran and V. Sasisekharan, *Advances in Protein Chemistry* (1968), **23**, 283. Reproduced by courtesy of the authors and publishers)

presumably provide further possibilities for free rotation. Some conformations are banned purely on geometrical grounds, for example, $\phi = \psi = 180°$ because in this situation the atoms O(A) and H(B) (Figure 7.1) would occupy almost exactly the same positions in space. Other conformations are energetically unfavourable. As with simpler molecules, there are overall preferred conformations which coincide with potential energy minima. Some of the many factors which contribute to the potential energy of the system are considered in the next few paragraphs.

The case discussed above of a theoretical conformation in which two atoms occupy the same space is an extreme one, but even when conformations cannot be completely excluded because of geometrical coincidence, atoms which are near to each other contribute to the overall potential energy by nonbonded interactions, At relatively long range, these interactions involve attractive (van der Waal's) forces, whilst at short range, due to the overlapping of the electron shells, the forces are repulsive. In addition, since all covalent bonds are to some extent polarized, electrostatic interactions between atoms (dipole–dipole interactions) influence the potential energy of the system. Hydrogen bonding, which is not well understood, but which is visualized as a similar type of interaction between an electronegative atom and a hydrogen atom, provides stronger attractive forces than van der Waal's interaction. In some situations, hydrogen bond formation between amide groups **(273)** is extremely important. Even less well understood are the so-called hydrophobic interactions of hydrocarbon side chains. These groups tend to be repelled by water so that in aqueous media, conformations are preferred in which the hydrophobic groups are clustered together to reduce their contact with water molecules. By the same token, polar groups tend to be found to the outside of such clusters. Ionic interactions between polar groups are sometimes important in particular situations, but they are not all that common.

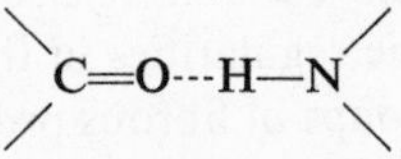

$$\rangle C{=}O\cdots H{-}N\langle$$

(273)

Normally, peptides, like simpler molecules, can be expected to adopt conformations which do not involve distortions of bond angles and bond lengths, but, taken together, the factors considered above can influence the molecule to adopt a conformation which involves some geometrical distortion. Obviously, there is an increase in the potential energy of the system as a result of this distortion and this has to be balanced by a net gain in stabilization energy. Such considerations can be important, for example, in cyclic peptides and in short sequences of amino acid residues joined by disulphide bridges.

In recent years, attempts have been made to take into account all of these various forces to calculate the most probable conformations (i.e. those of lowest potential energy) of a given peptide. Unfortunately, it is difficult to estimate in quantitative terms the influence of these forces and, at present, certain knowledge of preferred conformations is only provided by direct physical measurements. The most definitive studies have been made with proteins and there is still a grave shortage of techniques which provide information about the conformations of small peptides. That protein studies have been so successful is due, almost entirely, to x-ray diffraction techniques, which, in the first instance, provided an insight into the types of three dimensional atomic arrangements which occur in the fibrous proteins and which, more recently, have facilitated atom by atom analyses of crystalline globular proteins.

The conformations of fibrous proteins

Fibrous proteins often possess a regularity of structure which enables them, in a first approximation, to be regarded as polymers. Their X-ray diffraction patterns, which only possess 5–50 discrete deflections in all, as opposed to the 20–50 deflections per atom in a crystal diffraction pattern, can therefore be compared with the diffraction patterns predicted for specific model structures. Polymers prepared from amino acid *N*-carboxyanhydrides give X-ray diffraction patterns which resemble to a greater or lesser degree the patterns given by specific fibrous proteins and comparisons of this type are very helpful in the analysis. Ancillary information is derived from several sources, but particular mention must be made of polarized infrared absorption, which enables the relative directions of hydrogen bonds to be measured, and of electron microscopy, which can reveal the more macroscopic regularities in the fibre.

Classically, two major groups of fibrous proteins, the k-m-e-f group and the collagens, were distinguished on the basis of their distinctive X-ray diffraction patterns. The fibrous proteins of the k-m-e-f group can generally be stretched, whereas those of the collagen group are relatively inextensible. Included in the former are keratin, which is the structural material of nails, hair, horn and quills etc., myosin, which occurs in muscle, epidermin which occurs in the epidermis, and fibrinogen, the precursor of the protein fibrin, which participates in blood-clotting. Silk fibroin belongs to this group. Collagen is the major fibrous component of connective tissue. Many structural proteins have been investigated which do not strictly belong to either of these groups. Elastin, the protein of mammalian elastic fibres, resilin, the elastic protein of insect cuticle, and actin, which occurs in muscle, are three important examples. These proteins present structural problems in their own right, but the molecular arrangements encountered in the k-m-e-f and collagen groups have far-reaching significance. Basically, three types of structure are involved: the β-sheet and the α-helix of the k-m-e-f proteins, and the triple helix of collagen, although several other arrangements have been found.

The β-sheet, as exemplified by the stretched form of keratin, is characterized by repeating units (X-ray reflections) at 9·8 Å and 4·65 Å in equatorial positions and at 3·3 Å in the direction of the fibre axis (meridional position). This pattern is attributable to peptide chains which are held together laterally by hydrogen bonds at an average distance of 4·65 Å to form pleated sheets; the sheets are stacked vertically with an average distance between sheets of 9·8 Å (Figure 7.2).

β-Keratin has both parallel **(274)** and antiparallel **(275)** arrangements of the peptide chains. In silk fibroin only the parallel arrangement is found, and the β-sheet seems to be completely extended rather than pleated.

```
   /          /                 /              \
 OC         OC                OC                NH
   \      .'  \                 \              /
    NH          NH               NH------OC
   /           /                 /          \
 CHR    RHC                    CHR          CHR
   \           \                 \          /
    CO          CO               CO------HN
   /   `.      /                 /          \
 HN       HN                   HN            CO
   \        \                    \          /
    CHR      CHR                  CHR  RHC
   /        /                    /          \
 OC       OC                   OC            NH
   \    .'  \                    \          /
    NH        NH                  NH------OC
   /         /                   /          \

      (274)                          (275)
```

In the unstretched, α-form of keratin, the hydrogen bonds lie parallel to the axis of the fibre. x-Ray diffraction studies show that there is an equatorial deflection equivalent to a 9·5 Å repeat and one main meridional deflection which indicates a 1·5 Å unit. These findings indicate a helical structure (Figure 7.3) for α-keratin, in which the NH group of each

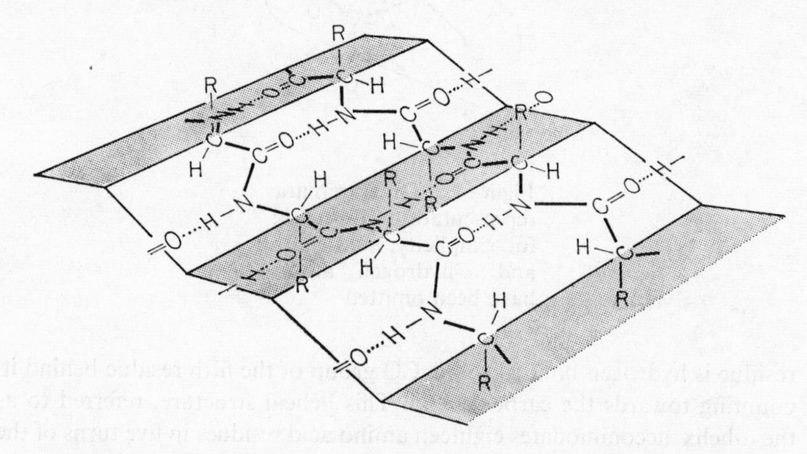

Figure 7.2. Diagrammatic representation of β-pleated sheet structure

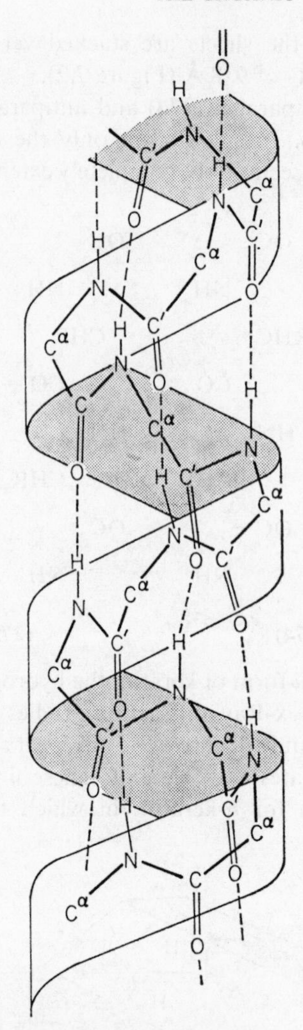

Figure 7.3. Diagrammatic representation of α-helix; for simplicity, side chain and α-hydrogen atoms have been omitted

residue is hydrogen bonded to the CO group of the fifth residue behind it, counting towards the carboxyl end. This helical structure, referred to as the α-helix, accommodates eighteen amino acid residues in five turns of the helix and is therefore said to be a $3 \cdot 6$ residue helix. Several other types of helix are known to occur in specific peptides and proteins, but the α-helix is the most stable and it occurs in many natural proteins and synthetic polyamino acids. Both right-handed and left-handed forms of the α-helix are known. In the right-handed form the side chains of the amino acid residues are further away from the carbonyl groups than in the left-handed form; it is the right-handed form which occurs in natural compounds. Helical chains in α-keratin occur in bundles, twisted together to form ropes and coiled coils, but the details of this coiling are not well understood.

Collagens possess a high glycine, proline and hydroxyproline content and their x-ray diffraction pattern is different from both β-sheet and α-helix patterns. In general terms, they are constructed of units of three peptide chains which are twisted about a common axis, each chain a helix with approximately 3·3 residues per turn (Figure 7.4).

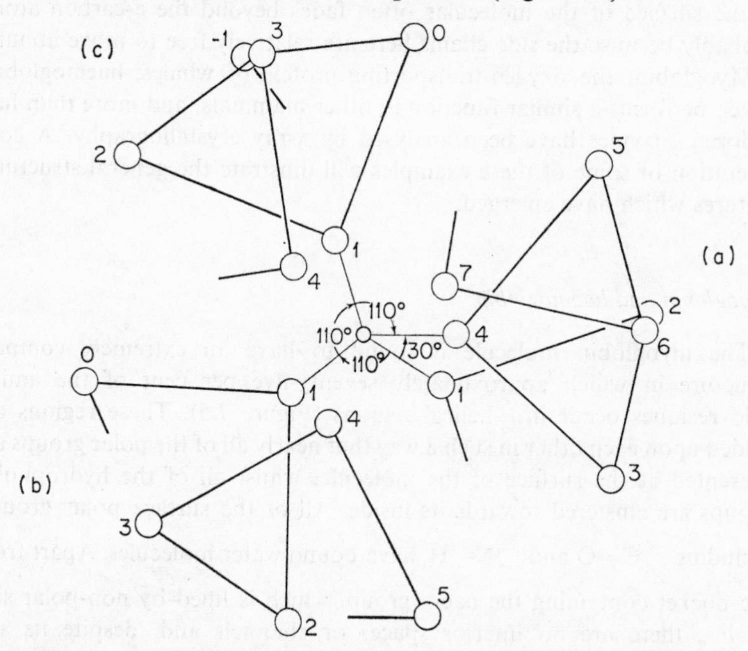

Figure 7.4. Schematic diagram of the triple chain coiled-coil structure of collagen showing the α-carbon atoms as circles and the peptide units as lines joining them. The three chains are marked (a), (b), (c) and the atoms with the same numerical symbols in the three chains are equivalent, as regards the helical coiling about the central axis. (From G. N. Ramachandran and V. Sasisekharan, *Advances in Protein Chemistry* (1968) **23**, 283. Reproduction by courtesy of the authors and publishers)

The conformations of globular proteins

The molecules of globular proteins, as the name implies, tend to be roughly spherical and devoid of the regular repeating units of the fibrous proteins. X-Ray diffraction methods can therefore only be used to determine the structures of globular proteins if the proteins can be obtained in a crystalline form, suitable for single crystal analysis. However, when a protein can be obtained in this form, a high resolution analysis can be carried out and the positions of each atom in the compound determined. At 2 Å resolution, the peptide chain can be traced accurately in a three dimensional electron density map. Side chains can usually be identified unequivocally, but since they are not completely resolved, sequence information derived from chemical studies is of inestimable help in confirming their identities. The electron density map of side chains which are at the surface of the molecular often fades beyond the β-carbon atom, probably because the side chains here are relatively free to move about.

Myoglobin, the oxygen-transporting protein of whales, haemoglobin, which performs a similar function in other mammals, and more than half a dozen enzymes have been analysed by X-ray crystallography. A consideration of some of these examples will illustrate the general structural features which have emerged.

Myoglobin and haemoglobin

The myoglobin molecule is found to have an extremely compact structure in which approximately seventy-five per cent of the amino acid residues occur in α-helical regions (Figure 7.5). These regions are folded upon each other in such a way that nearly all of the polar groups are presented at the surface of the molecule, whilst all of the hydrophobic groups are clustered towards its inside. All of the surface polar groups, including $\rangle$C=O and $\rangle$N—H, have bound water molecules. Apart from the pocket containing the haem group, which is lined by non-polar side chains, there are no interior spaces or channels and, despite its size (M = 17,000), only four or five molecules of water of crystallization are included in the molecule. There is almost complete van der Waal's contact between side chains and most of the stabilization energy of the tertiary structure seems to be derived from van der Waal's forces, rather than from polar interactions or intrachain hydrogen bonding. The characteristics of the helical regions of myoglobin closely resemble those of the α-helix in α-keratin. As usual, the right-handed form of the helix is involved.

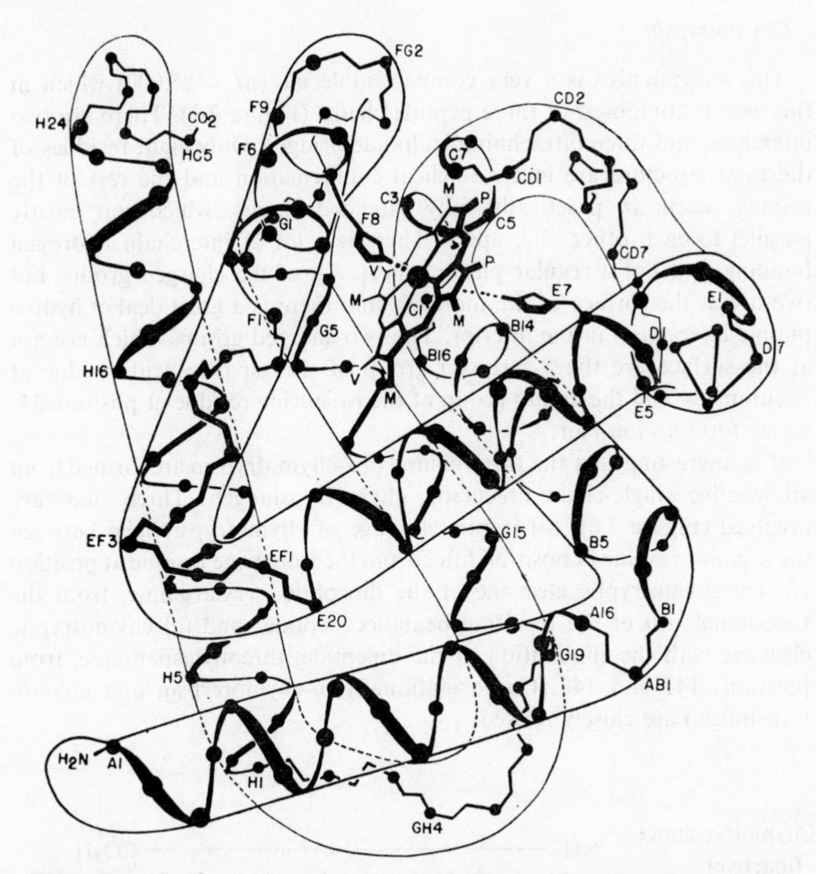

Figure 7.5. Conformation of sperm-whale myoglobin. The haem moiety–iron bearing tetrapyrrole (porphyrin) structure—is seen coordinated between two histidine residues; all other side chains have been omitted for simplicity (after R. E. Dickerson, in *The Proteins*, Ed. H. Neurath, Academic Press (1964) **2**, 634. Reproduced by courtesy of the author and publishers)

Thus, the myoglobin molecule which might, at first sight, seem rather a jumble, is energetically highly favoured. Indeed, it seems likely that the energetic preference for this conformation is so great that in solution it will be virtually the same as in the crystalline form. Haemoglobin is remarkably similar and at 5·5 Å resolution, the convolutions of the peptide chains of myoglobin and haemoglobin are almost indistinguishable. However, the haemoglobin molecule (M = 67,000) possesses four peptide chains, each arranged so that it is in van der Waal's contact with two others.

α-Chymotrypsin

This enzyme also is a very compact molecule ($M = 25{,}000$) which in this case is composed of three peptide chains (Figure 7.7). There are two interchain and three intrachain disulphide bridges. Only eight residues of the total structure are in an α-helical conformation and the rest of the residues occur in practically fully extended chains, which run mostly parallel to each other, 5 Å apart. There is a lot of interchain hydrogen bonding, but not a regular pleated sheet. All of the charged groups but two are at the surface of the molecule and there is a great deal of hydrophobic interaction in the interior. The two charged groups which are not at the surface are the β-carboxyl group of the aspartic acid residue at position 194 and the amino group of the isoleucine residue at position 16, which form an ion pair.

It is interesting that the three chains of α-chymotrypsin are formed from an inactive single-chain precursor, chymotrypsinogen. Three steps are involved (Figure 7.6): (a) tryptic cleavage of chymotrypsinogen between the arginine residue at position fifteen and the isoleucine residue at position 16; (b) chymotryptic cleavage of the dipeptide, serylarginine, from the *C*-terminal end of the resulting pentadecapeptide; and (c) chymotryptic cleavage with the elimination of the dipeptide, threonylasparagine, from positions 147 and 148. Conformationally, α-chymotrypsin and chymotrypsinogen are closely related.

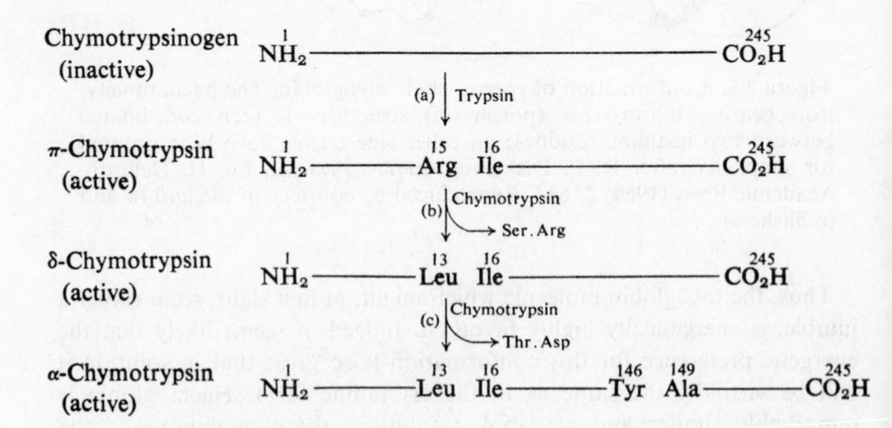

Figure 7.6. Diagrammatic representation of the formation of the three peptide chains of α-chymotrypsin (after L. Stryer, *Annual Reviews of Biochemistry*, 1968, 37, 29. Reproduced by courtesy of the author and publishers)

Carboxypeptidase A

Approximately thirty per cent of the amino acid residues in carboxypeptidase A ($M = 34{,}600$) are in helical regions (Figure 7.8) and these are confined to the outside of the molecule; some twenty per cent of the residues are in fully extended peptide chains which form a large, central twisted sheet with four pairs of parallel and three pairs of antiparallel portions; the rest of the amino acid residues are located in an elaborately folded coil region. The primary structure of this enzyme has not been determined by chemical methods but the part sequences which have been determined in this way agree with the X-ray findings.

Lysozyme

The enzyme lysozyme ($M = 14{,}600$), which degrades the mucopolysaccharide of bacterial cell walls (p. 177), consists of a single peptide chain which possess four intrachain disulphide bonds (Figure 7.9). Just under half of its one hundred and twenty-nine amino acid residues occur in slightly distorted α-helical regions. Two lengths of chain adopt an antiparallel pleated sheet arrangement. Most of the hydrophobic side chains of lysozyme are clustered within the molecule and all of the polar groups are at the surface.

Ribonuclease A

This enzyme ($M = 14{,}000$) catalyses the hydrolysis of ribonucleic acids (Chapter 8). Its primary structure, which consists of a single peptide chain with four disulphide bonds, has already been considered (p. 125). Approximately fifteen per cent of its residues are located in a helical region towards the *N*-terminal end of the molecule (Figure 7.10) and a hydrophobic cluster, through which the main chain does not run, accounts for a further fifteen per cent of the molecule. Otherwise, the peptide chain is folded into roughly antiparallel regions. All of the charged side chains of ribonuclease are on the surface of the molecule with the exception of the aspartic acid residue at position 14 and the histidine residue at position 48.

Peptide conformation: general conclusions

Proline cannot be accommodated in an α-helix and valine and isoleucine are often found in non-helical regions. Otherwise, one of the most surprising findings of the X-ray crystallographic studies is that there is no

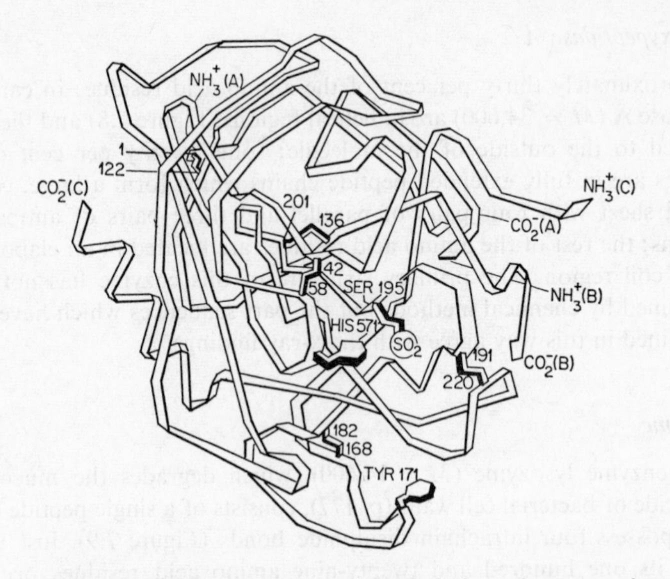

Figure 7.7. Folding of the protein chain in α-chymotrypsin (reproduced by courtesy of Dr D. M. Blow and the Medical Research Council)

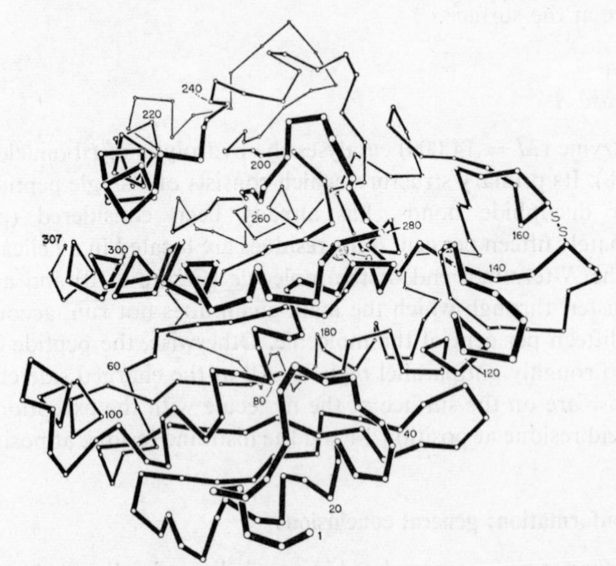

Figure 7.8. Folding of the protein chain in carboxypeptidase A (reproduced by courtesy of Professor W. N. Lipscomb)

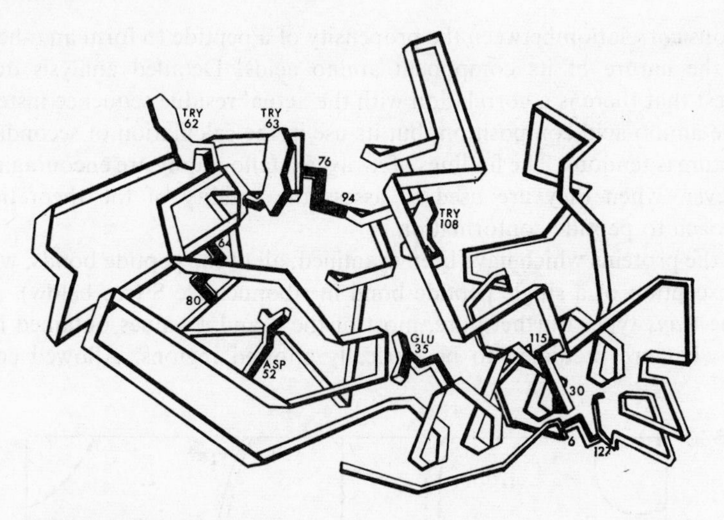

Figure 7.9. Folding of the protein chain in lysozyme (C. C. F. Blake, *Wallerstein Laboratories Communications*, (1968) **31**, 127. Reproduced by courtesy of the author and publishers)

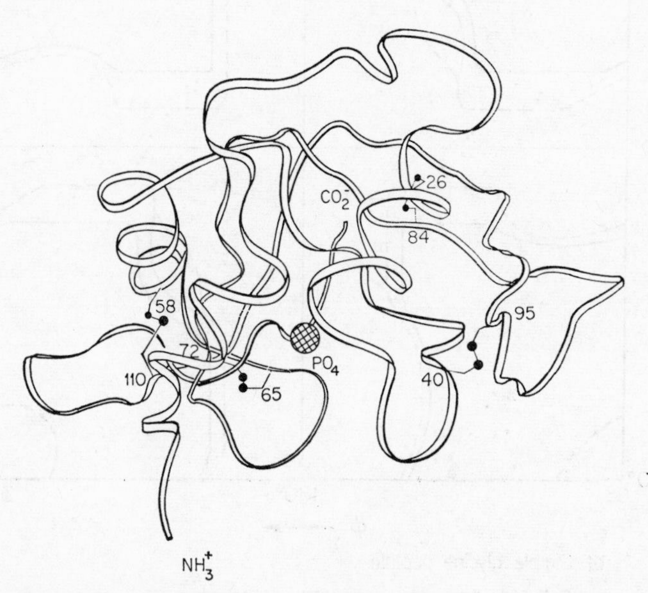

Figure 7.10. Folding of the protein chain in ribonuclease (G. Kartha, *Chemical Research* (December, 1968) **1**, 379. Reproduced by courtesy of the author and of the publishers, the American Chemical Society)

obvious correlation between the propensity of a peptide to form an α-helix and the nature of its component amino acids. Detailed analysis does suggest that there is a correlation with the actual residue sequence instead of the amino acid composition, but its use in the calculation of secondary structure is tenuous. The findings of X-ray crystallography are encouraging, however, when they are used to assess the validity of the theoretical approach to peptide conformation.

In the proteins which have been examined, all of the peptide bonds, with the exception of a single peptide bond in ribonuclease S (see below), are of the *trans* type. Furthermore, most of the ϕ and ψ values obtained fall into what are predicted to be sterically allowed regions. Allowed con-

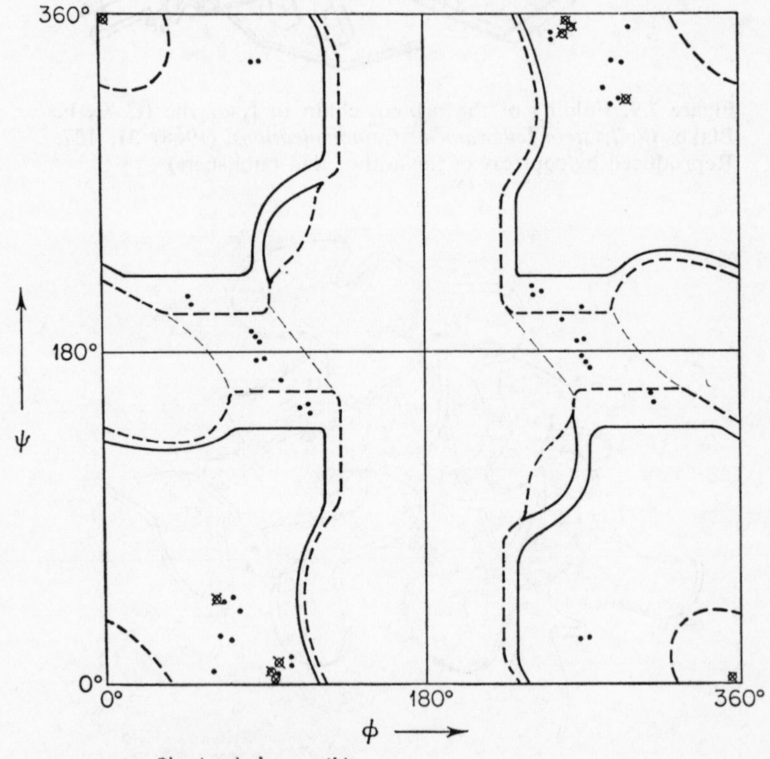

Figure 7.11a. Conformational map for peptides in which β-carbon atoms are absent, $\tau = 110°$. Fully allowed conformations are outlined by a solid line and partially allowed conformations by a broken line

formations can be described conveniently by plotting the relevant potential energy isobars between reference axes of ϕ and ψ to give what is termed a conformational map. The angle at the α-carbon atom of the peptide backbone (τ = angle $NC^{\alpha}C'$, Figure 7.1) is defined for each map. When no β-carbon atoms are involved, i.e. with glycine peptides, such a map ($\tau = 110°$) reveals that forty-five per cent of the area is fully allowed; another sixteen per cent is available under extreme circumstances (Figure 7.11a). By contrast, when a β-carbon atom is present, as in peptide units

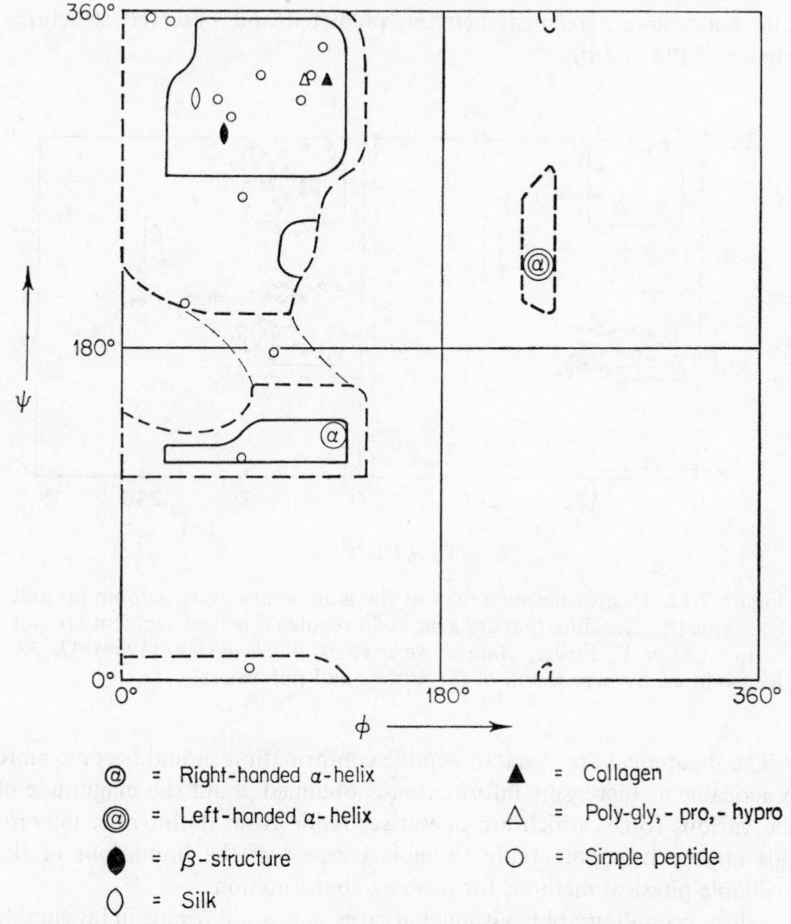

Figure 7.11b. Conformational map for peptides in which the β-carbon atoms are present. (Both Figures 7.11 a and b are from G. N. Ramachandran and V. Sasisekharan, *Advances in Protein Chemistry* (1968) **23**, 283. Reproduced by permission of the authors and publishers)

composed of alanine residues, only seven and a half per cent of the area is fully allowed and a further fifteen per cent falls within the extreme limits (Figure 7.11b). In either case, increasing τ to 115° slightly increases the allowed area, whereas decreasing τ to 105°, reduces it. These values are likely to represent the extremes which can be encountered. Branching beyond the β-carbon atom hardly affects the conformational map.

Because of the relative uncertainty about the orientations of the amino acid side chains, only approximate ϕ and ψ values can be measured from the models produced from the X-ray diffraction data, but even so there is quite reasonable agreement between predicted and observed structures (Figures 7.12a and b).

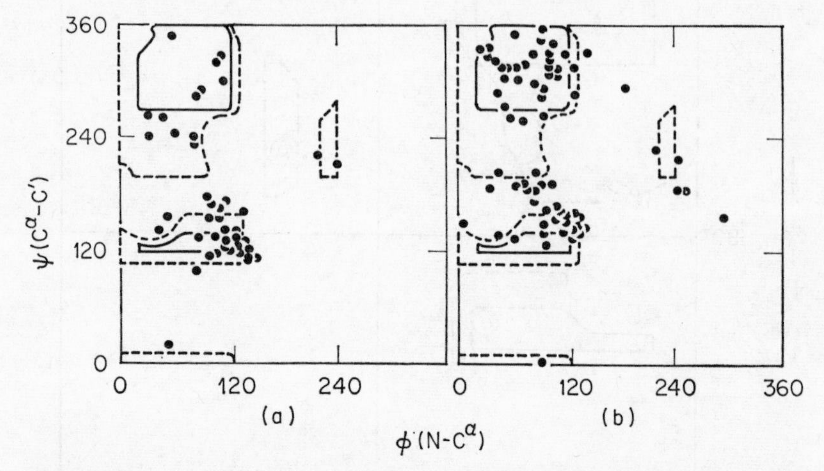

Figure 7.12. Conformational map of the main chain in myoglobin (a) and lysozyme (b). Residues that are a part of a regular α-helical segment are not shown (After L. Stryer, *Annual Reviews of Biochemistry*, (1968) **37**, 38. Reproduced by permission of the author and publishers)

The theoretical approach to peptide conformation should become more practicable as increasing information is obtained about the magnitude of the various forces which are operative. Aside from its intrinsic interest, this approach is potentially valuable because of the limitations of the available physical methods for probing conformation.

X-Ray crystallography, although it gives such good results in favourable instances, is not of universal applicability. For large proteins there is the difficulty that X-ray crystallographic techniques can only be applied to crystals with unit cell dimensions of less than 300 Å, but more often than not, the problem of crystallizing the substance at all is a greater stumbling block. *A priori*, before a compound can be studied by crystallography, it has to be obtained in a crystalline form. In addition, it is generally necessary at present to attach one or more heavy atoms to the molecule to facilitate the X-ray analysis and there is always the danger that this interference with the parent molecule might cause conformational changes. In the peptide field, where crystallinity is elusive, these requirements are often prohibitive. For this reason, at the time of writing, none of the biologically active small peptides which have been discussed in previous chapters have been investigated by this technique.

No technique of comparable incisiveness to X-ray crystallography exists for peptides which will not crystallize. The conformational models which are available for such compounds are due to a combination of physical techniques, including ultraviolet and infrared spectroscopy, nuclear magnetic resonance spectroscopy, optical rotatory dispersion (ORD), circular dichroism (CD), dialysis and surface studies. So far as is known, ORD and CD provide a reasonable indication of the helical content of peptides and it is a source of comfort that many of the predictions made by applying these techniques to macromolecules have been vindicated by subsequent X-ray crystallographic studies. However, none of the conformational models yet proposed for small peptides satisfy all of the available data.

The biological roles of peptides related to their structures

Superficially, it might be thought that peptides fulfil three different types of biological role. One of these is dependent on physical properties and the main compounds concerned are structural proteins; another involves physicochemical properties which can usually be demonstrated outside the biological system, and the compounds concerned are globular proteins; the third depends on properties which are only manifest within the context of the biological system and the compounds concerned are hormones, antibiotics, toxins etc. This classification is neither comprehensive nor rigorous but it should provide a useful framework for a discussion which attempts to correlate the structures of peptides with their biological roles.

Structure proteins

The physiological demands made upon the structure proteins can be defined fairly accurately in mechanical terms. In various situations they must form fibres or extended sheets which will imbue a tissue with rigidity, elasticity, ability to expand and contract, hardness, tensile strength etc. It is rare that the structure proteins occur alone. Frequently, they are associated with other compounds, perhaps with carbohydrates or globular proteins, and act as components of a highly organized biological system. For example, the protein, resilin, with the polysaccharide, chitin, forms a rubber-like lamellated structure which is responsible for the elastic return of insects' wings during flight. It is not always clear what should be expected of the protein component after it has been extracted from such complex structures, nor is it always clear how the individual units of such a composite structure fit together and the rationalization of such phenomena in molecular terms goes beyond the realm of organic chemistry.

Even when the constituent protein has been separated from the biological system it is still a complex structure. The characteristic X-ray diffraction patterns on which the α-helix and other structures are based, are usually only obtained from certain regions of the protein mass. These regions are separated by areas in which the peptide chains seem to be irregularly arranged. Probably, it is the interplay of these two strikingly different regions which accounts for the behaviour of an individual chain. It must also be remembered that the protein will generally be composed of a large number of these chains. The twisting and coiling of the chains, their lateral assembly to form sheets, the movement of one chain relative to another or of one sheet relative to another, the making and breaking of various types of cross linkages, including disulphide bonds, have all been invoked to account for specific physical functions. There is ample scope for the conformations discussed earlier in this chapter to participate in these transformations.

The elastic proteins, resilin and elastin, are of particular interest because they seem quite devoid of regular structure. Furthermore, in terms of amino acid analysis and therefore of primary structure, they are remarkably different, although they perform similar functions in insects and in vertebrates respectively. Both, when hydrated, are almost ideal protein rubbers. They become brittle and glass-like when dried, but the rubberiness can be restored by wetting them. They differ therefore from real rubber in not being self-lubricating. The solvation which is essential for their elasticity probably involves hydrogen bonding of water to the peptide backbone. Both proteins apparently consist of a three dimensional network of cross-linked, randomly coiled, peptide chains and qualitatively, it is tempting to attribute their elastic properties to this molecular organisation. Neither protein contains cystine and the elucidation of the nature of the cross-linking between the chains has been difficult. In both proteins unusual amino acids are involved. A tyrosine derivative (**276**) is thought to occur in resilin, whereas pyridinium structures (**277**, **278**; $R^1 = (CH_2)_2.CH(NH_2).CO_2H$; $R^2 = (CH_2)_4.CH(NH_2).CO_2H$; $R^3 = (CH_2)_3.CH(NH_2).CO_2H$), derived from lysine, account for the cross-linking in elastin.

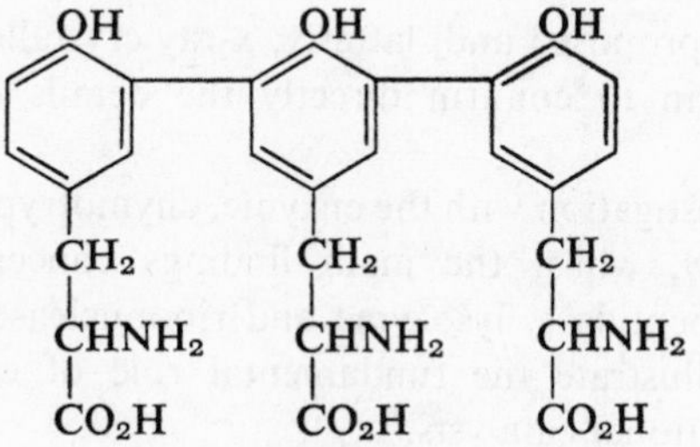

(276)

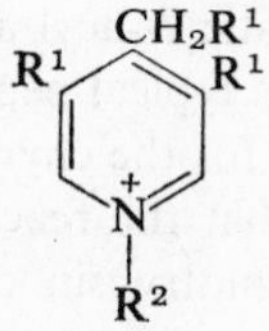

(277) 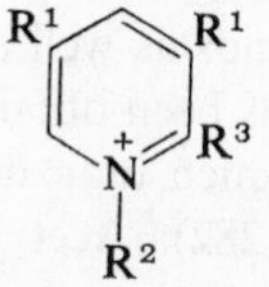 **(278)**

Globular proteins

The functions of the globular proteins are manifold and many of them have not been clearly defined. Plasma proteins in vertebrates, for example, are responsible for the maintenance of osmotic pressure, pH and electrolyte balance, and for the transport of various substances including oxygen and metal ions. They participate in blood clotting and in the immune response, and they both transport and give rise to various hormones and other pharmacologically active substances. In addition, many enzymes circulate in the plasma, as normal constituents or as a result of cellular damage somewhere in the organism. Attention here will be focussed on enzymic reactions both because of their supreme importance and because, at the molecular level, more is known about enzymic reactions than about any other type of peptide function.

It has been understood for many years that enzymic reactions generally take place in three or more kinetically independent stages. In the first place the enzyme and substrate form a complex in which reactive groups in the enzyme molecule are favourably situated to react with the substrate.

Subsequently, these groups react with the substrate molecule to cleave it and, in the process, a part of the substrate molecule becomes covalently bound to the enzyme. Finally, the covalently-bound fragment of the substrate is detached leaving the enzyme molecule free to complex with another molecule of substrate. Each stage of the reaction is reversible. Chemical studies with specific enzymes have enabled particular mechanisms to be proposed and, latterly, X-ray crystallography has made it possible to begin to confirm directly the details of some of these reactions.

This type of investigation with the enzyme, chymotrypsin, is discussed in some detail below, whilst the main findings concerning three other enzymes, carboxypeptidase, lysozyme and ribonuclease are summarized. These examples illustrate the fundamental role of enzymes as highly selective multifunctional catalysts.

α-Chymotrypsin

This enzyme cleaves esters of aromatic *N*-acyl amino acids and esters of simple carboxylic acids as well as its typical peptidic substrate (p. 17). Kinetic information has been obtained for the chymotryptic hydrolysis of *p*-nitrophenylacetate which indicates that the reaction proceeds in three discrete steps (**279** → **282**). Acetylchymotrypsin can be isolated as an intermediate of the reaction.

$$\text{enzyme} + CH_3.CO.O.C_6H_4.NO_2(p) \quad \textbf{(279)}$$

$$\rightleftharpoons$$

$$\left[\text{enzyme} - CH_3.CO.O.C_6H_4.NO_2 \atop \text{Complex}\right] \quad \textbf{(280)} \rightleftharpoons CH_3\ CO.\text{enzyme} + HO.C_6H_4.NO_2(p) \quad \text{(acyl enzyme)} \quad \textbf{(281)}$$

$$\rightleftharpoons$$

$$CH_3.CO_2H + \text{enzyme} \quad \textbf{(282)}$$

Both the acylation and deacylation steps of chymotryptic hydrolysis are influenced by a group which possesses a pK_a of 7. It is postulated that this group is the imidazole moiety of a histidine side chain and this is substantiated by studies in which the enzyme was reacted with a chloromethyl ketone (**283**) related to phenylalanine. This substance is recognized by the enzyme, despite the fact that it lacks an appropriate hydrolysable bond, and a substituted alkyl derivative of the enzyme results. Degradation reveals that it is the imidazole ring of the histidine residue in position 57 which is alkylated. The alkylated enzyme is completely inactive.

$$(p)CH_3.C_6H_4.SO_2.NH.\underset{}{\overset{CH_2Ph}{\overset{|}{C}}}H.CO.CH_2Cl$$

(283)

α-Chymotrypsin can also be inactivated with diisopropylphosphorofluoridate **(284)**. In this instance, a phosphate ester of the hydroxymethyl side chain of the serine residue in position 195 is formed. This serine residue is unusually reactive because, although there are thirty serine residues in the molecule, it is always the same one which forms the diisopropylphosphate. The sequence of the residues around the reactive serine has been determined by partial degradation studies. It is found that several enzymes which have similar functions, including trypsin, elastase, cholinesterase, thrombin, and alkaline phosphatase, possess remarkably similar amino acid sequences about a reactive serine residue. In all cases, an acidic amino acid residue, either aspartic acid or glutamic acid, occurs immediately before the serine residue.

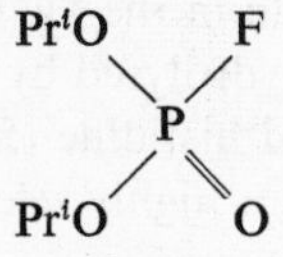

$$(Pr^iO)_2P(=O)F$$

(284)

The same serine hydroxyl group is involved in the acetylchymotrypsin derivative isolated from the *p*-nitrophenylacetate studies. It is also implicated in the deactivation of chymotrypsin with *p*-toluene sulphonyl chloride, because the *O*-tosylate which forms can be converted by treatment with base to give a dehydroalanine residue at this position (p. 53).

Acetylation of the α-amino group of the isoleucine residue in position 16 deactivates the enzyme, although all other amino groups in the molecule can be acetylated without loss of activity. The importance of this amino group is further indicated by the pH-dependence of the formation of the enzyme–substrate complex. Complex formation is controlled by a group which has a pK_a value of approximately 8·5. In fact, it only occurs when the amino group of the isoleucine residue is protonated.

A special role must be attributed to the methionine residue at position 192 in α-chymotrypsin. When the enzyme is treated with hydrogen peroxide, this residue is converted to the corresponding sulphoxide. The activity of the oxidized enzyme is very much reduced and, again, it is the first stage in the reaction, the formation of the enzyme–substrate complex, which is affected. This suggests that the methionine residue and its immediate environment are concerned with the initial recognition and binding of the substrate.

In summary, chemical studies suggest that the enzymic reactivity of chymotrypsin specifically involves the following residues; serine (195), histidine (57), isoleucine (16), methionine (192) and possibly, since an acidic residue always occurs in this position, aspartic acid (194).

From the X-ray crystal structure, it is clear that the interaction of the amino group of isoleucine (16) with the carboxyl group of aspartic acid (194) has a profound effect on conformation, particularly since both residues are hydrogen-bonded to other residues. It will be recalled that the activation process involves the liberation of the amino group of isoleucine (16) in the chymotrypsinogen precursor. A limited conformational change must accompany this step and it has been suggested that a loop of peptide chain swings out over an indentation to make a pocket in the molecule.

It has been proposed that the stereochemistry of the resulting molecule is such that complex formation can occur by hydrophobic binding of the aromatic moiety of the substrate in the vicinity of methionine (192), and that the adsorbed substrate is hydrolysed by general acid–base catalysis in which both the serine (195) and histidine (57) residues play a part. Thus the enzyme acts (**285** → **288**) as a highly selective bifunctional catalyst.

Carboxypeptidase A

The availability of X-ray pictures, both of the native enzyme and of an enzyme–substrate complex, makes it possible to visualize the conformational changes associated with carboxypeptidase catalysis. When glycyltyrosine combines with the enzyme to form such a complex, there is an appreciable movement of a tyrosine side chain in the protein. This movement involves rotation about the C^{α}–C^{β} bond of the side chain, together with some movement of the backbone, and it effectively brings the phenolic group close to the susceptible peptide bond of the substrate. The carbonyl group of this amide is probably coordinated to the zinc atom of the enzyme; its carboxyl group interacts with the guanidino group of an arginine side chain in a way which involves some movement of the arginine side chain. It has been proposed that the carboxyl group of an aspartic or glutamic acid residue forms a transient mixed anhydride with the carbonyl group of the peptide and that the tyrosine phenolic group donates its proton to the nitrogen atom of the peptide bond. The importance of the

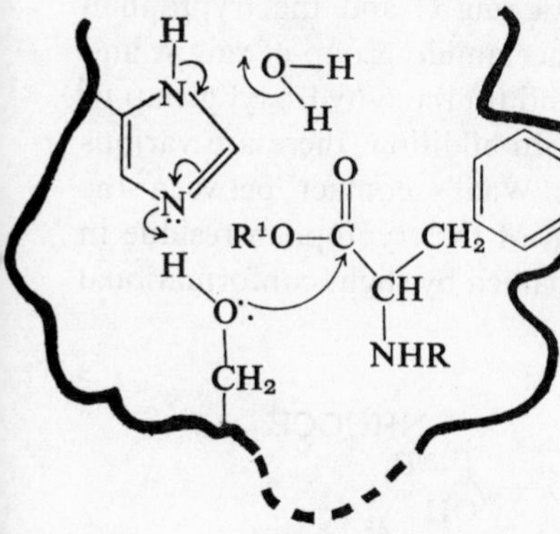

(285) 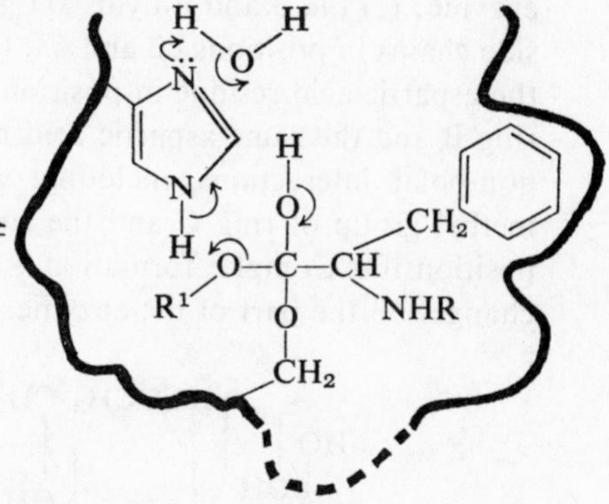 ⇌ (286)

(286) ⇌ R^1OH + (287) ⇌ enzyme + RNH.CH(CH_2Ph).CO_2H + R^1OH (288)

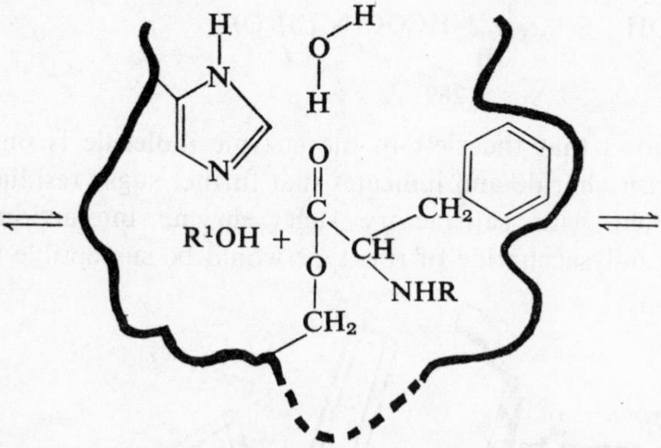

zinc atom in this scheme is that it exerts a fundamental influence on the overall stereochemistry and activates the carbonyl group.

Lysozyme

The *N*-acetylglucosamine trisaccharide (**289**), which is related to the mucopolysaccharide of bacterial cell walls, forms a specific complex with the enzyme, lysozyme, but is not cleaved by it. Presumably, the sugar molecule is too small to be digested. x-Ray crystallographic studies of the complex show exactly how the trisaccharide is bound to the enzyme and have enabled a detailed mechanism to be proposed for the mode of action of the enzyme. The trisaccharide is held in a cleft in the enzyme by hydrogen bonding between (a) the carbonyl group of the acetamido side chain of ring **C** and the backbone NH of residue 59 of the enzyme (b) the NH of the acetamido group of ring **C** and the backbone CO of residue 107 of the enzyme; (c) the 3 and 4-hydroxyl groups of ring **C** and the tryptophan side chains in positions 62 and 63; (d) the acetamido group of ring **A** and the aspartic acid residue in position 101; and (e) the 6-hydroxyl group of ring **B** and the same aspartic acid residue. In addition, there are various non-polar interactions, including van der Waal's contact between the methyl group of ring **C** and the side chain of the tryptophan residue in position 108. Complex formation is accompanied by slight conformational changes on the part of the enzyme.

NHCOCH$_3$ CH$_2$OH NHCOCH$_3$
HO OH O OH O OH OH
CH$_2$OH NHCOCH$_3$ CH$_2$OH
A **B** **C**

(289)

Model building shows that the cleft in the enzyme molecule is only partly filled by the trisaccharide and indicates that further sugar residues can be accommodated with satisfactory sugar–enzyme interactions. Presumably, a larger polysaccharide of this type would be susceptible to

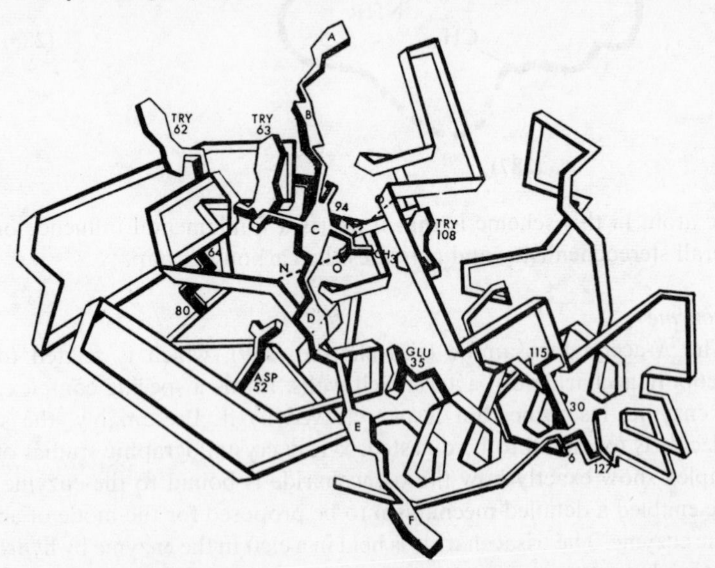

Figure 7.13. Lysozyme–substrate complex (C. C. F. Blake, *Wallerstein Laboratories Communications* (1968) **31**, 127. Reproduced by courtesy of the author and publishers)

enzymic cleavage and the model gives an indication of how the cleavage might occur. To accommodate a fourth ring (**D**) of a polysaccharide in the complex, it is necessary to distort it from its customary chair conformation so that the oxygen atom and carbon atoms 1, 2 and 5 lie in a plane. Model building also leads to the inescapable conclusion that it is the bond between the C_1 carbon atom of this ring and the oxygen atom derived from the 4-hydroxyl group of the next sugar residue which is broken during lysozymic digestion. The side chains of aspartic acid and glutamic acid residues at positions 52 and 35 respectively seem to participate in the catalysis (Figure 7.13). It is postulated that the glycosidic oxygen atom is protonated by the glutamic acid side chain and that cleavage of the C_1–O bond follows with the formation of a carbonium ion at C_1 (**290** → **291**). The carbonium ion is stabilized due to the planarity of the ring and through its electronic interaction with the side chain of the aspartic acid residue. Finally, the carbonium ion is hydroxylated by the aqueous medium.

Ribonuclease

x-Ray crystallographic studies of this enzyme suggest that its active site includes the residue sequences in the vicinity of histidine 12, histidine 48 and histidine 119, all of which lie fairly close together in a depression in the side of the kidney-shaped molecule. Chemical studies also indicate

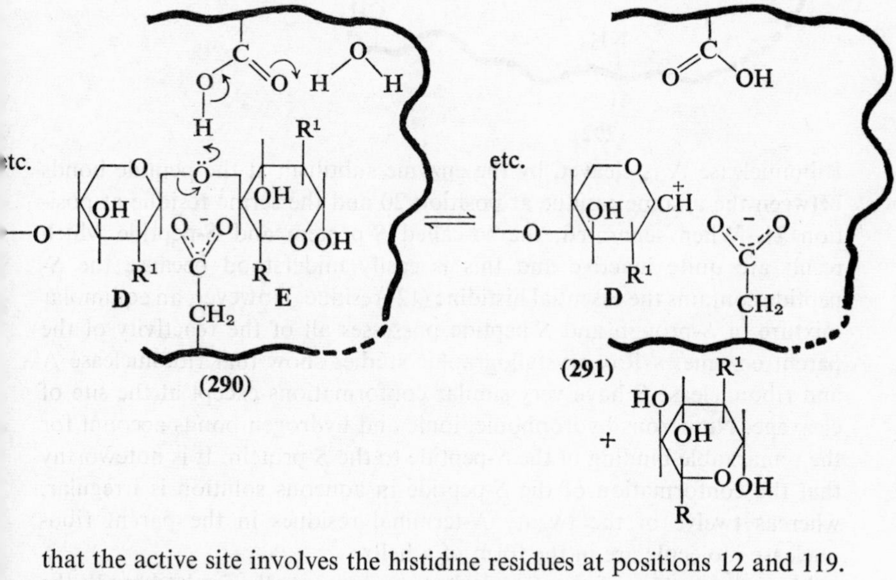

(**290**) ⇌ (**291**) + [product sugar: HO, OH, O OH, R^1, R]

that the active site involves the histidine residues at positions 12 and 119. It seems likely that the formation of the enzyme–substrate complex in this case depends upon (a) the interaction of the phosphate group of the nucleotide (Chapter 8) with the ϵ-amino group of the lysine residue in position 41; and (b) the hydrogen bonding of the pyrimidine ring of the nucleotide, via a water molecule, to one of the histidine residues (24). It is proposed that the conformational change thus induced in the enzyme is sufficient to bring the imidazole ring of the other histidine residue within hydrogen-bonding distance of the 2-hydroxyl group of the ribose moiety, thereby facilitating 2,3-cyclic phosphate formation and cleavage of the substrate (**292** → **293**).

$\rightleftharpoons$ ROH +

(**292**) (**293**)

Ribonuclease A is cleaved by the enzyme subtilisin at the peptide bonds between the alanine residue at position 20 and the serine residue at position 21. When separated, the so-called *S*-protein and *S*-peptide which result are quite inactive and this is easily understood because the *S*-peptide contains the essential histidine (12) residue. However, an equimolar mixture of *S*-protein and *S*-peptide possesses all of the reactivity of the parent enzyme. x-Ray crystallographic studies show that ribonuclease A and ribonuclease S have very similar conformations except at the site of cleavage. Numerous hydrophobic, ionic and hydrogen bonds account for the remarkable binding of the *S*-peptide to the *S*-protein. It is noteworthy that the conformation of the *S*-peptide in aqueous solution is irregular, whereas twelve of the twenty *N*-terminal residues in the parent ribonuclease molecule are in the form of a helix.

Many analogues of the *S*-peptide have been synthesized to study the effect of modifying different side chains. It proves that the glutamic acid residue in position 2, aspartic acid residue in position 14 and methionine residue in position 13, participate in the binding of *S*-protein to *S*-peptide. These residues are not essential for activity although they do exert an effect. On the other hand, synthetic analogues confirm that the histidine residue in position 12 is essential for activity. If this residue is replaced by β-(pyrazol-3-yl)-alanine (**294**), which is sterically similar to histidine but which has very different acid–base properties, the peptide which results does not activate *S*-protein even when present in a molar ratio of 1000:1. On the other hand, it does competitively inhibit *S*-peptide activation which shows that it is bound by the protein.

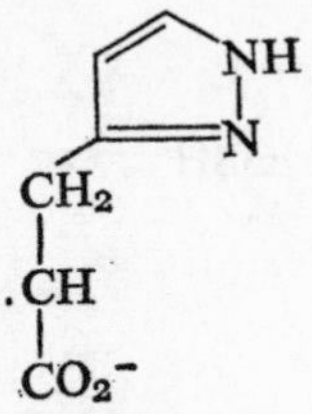

NH
N
CH_2
$\overset{+}{N}H_3.CH$
CO_2^-
(**294**)

*Exercise 7.4**†

Mandelic acid (**295**) is formed when 2-(*N,N*-dimethylamino)-ethyl mercaptan is added at room temperature to an aqueous solution of phenyl glyoxal (**296**). Mixtures of individual thiols and amines are much less effective in this reaction, whilst both quarternary salts and thioethers of the above catalyst are inactive. The hydrogen atom attached to the benzyl carbon atom in the mandelic acid does not originate from the aqueous medium.

$Ph.CHOH.CO_2H.$ (**295**) $Ph.CO.CHO.$ (**296**)

Suggest a mechanism for this reaction.

† See Appendix.

Small peptides: hormones, antibiotics, toxins etc.

Glutathione functions as a coenzyme in the conversion of methylglyoxal (**297**) into lactic acid (**298**). The first stage of this reaction involves the formation of a hemithioacetal (**299**) which is converted by the enzyme glyoxylase I into the glutathione thiol ester of lactic acid (**300**). This conversion involves intramolecular hydride ion transfer in a manner rather reminiscent of the Cannizzaro reaction. It is inhibited by compounds which resemble glutathione but which lack a thiol group, for example, by ophthalmic acid and norophthalmic acid. Presumably, the reaction is best formulated as a type of bifunctional catalysis and on this basis it can be simulated by model compounds (Exercise 7.4*). It can therefore be expected that glyoxylase I will have a basic group at its active site. The final stage of the reaction involves the hydrolysis of the thiol ester by the enzyme, glyoxylase II.

$$\underset{\mathbf{(297)}}{\mathrm{Me.CO.CHO}} + \mathrm{GSH} \rightleftharpoons \mathrm{Me.}\overset{\mathrm{O}}{\overset{\|}{\mathrm{C}}}-\overset{\mathrm{OH}}{\overset{|}{\underset{\underset{\mathrm{H}}{|}}{\mathrm{C}}}}-\mathrm{SG}$$

(299)

$$\mathbf{(299)} \rightleftharpoons \mathrm{Me.}\overset{\mathrm{OH}}{\overset{|}{\underset{\underset{\mathrm{H}}{|}}{\mathrm{C}}}}-\overset{\mathrm{O}}{\overset{\|}{\mathrm{C}}}-\mathrm{SG}$$

(300)

$$\mathbf{(300)} \overset{\mathrm{H_2O}}{\rightleftharpoons} \underset{\mathbf{(298)}}{\mathrm{Me.}\overset{\mathrm{OH}}{\overset{|}{\underset{\underset{\mathrm{H}}{|}}{\mathrm{C}}}}\mathrm{.CO_2H}} + \mathrm{GSH}$$

[GSH = glutathione]

At least two more enzymic reactions also require the participation of glutathione as a coenzyme and it is possible that other systems may be dependent upon it. Glutathione, in concentrations as small as 10^{-5} molar, promotes a typical feeding reaction in hydra (*Hydra littoralis*) and it is likely that this has physiological significance since hydra always feeds on fresh tissues which must contain glutathione. The aspartic acid analogue of glutathione does not elicit this response which shows that the reaction is fairly specific. Opththalmic acid does give a positive response so that, for this particular activity, the thiol group of glutathione is not essential. In this system glutathione seems to perform a hormonal role.

Peptide hormones, unlike enzymes, which only require the presence of the relevant molecules for them to exhibit activity, manifest their effects in biological systems of a high order of complexity. Generally, more than one effect is observable. The pituitary hormones, oxytocin and vasopressin illustrate this point. Oxytocin can bring about contractions of the uterus (oxytocic effect), as in labour, and contractions of the myoepithelial tissue of the lactating mammary gland to cause milk ejection (milk let-down effect); in birds, its administration leads to a decrease in blood pressure (avian depressor effect). Vasopressin stimulates the concentration of urine in the kidney (antidiuretic effect) and causes an increase in blood pressure in rats (pressor effect). Simpler systems can often be devised for the detection and measurement of hormones, but the relationship between these artificial tests and the situation which prevails in the intact organ or organism is often uncertain. For example, oxytocin and its analogues stimulate, to varying degrees, the active transport of sodium ions through frog skin, but the potency of different compounds in this test does not parallel exactly their potency with respect to the types of activity discussed above. The same applies to the relatively simple observation that the permeability of toad bladder to urea is increased by the addition of vasopressin analogues.

When a compound is injected into a whole organism or tissue the observed effect is not a direct indication of how that compound reacts at the receptor responsible for the effect. Many other factors, such as the mobility of the compound within the system, its susceptibility or resistance to enzymic degradation, and its rate of excretion, must be taken into account. Individually these factors might tend either to enhance or to diminish the observed response.

Peptide hormones are remarkable for their potency. Often, fantastically minute amounts of them are sufficient to bring about a biological response. For example, in man, $2{\cdot}5 \times 10^{-6}$ milligrams of arginine–vasopressin are sufficient to bring about a measurable antidiuretic effect, whilst an intravenous drip containing 2×10^{-6} milligrams of oxytocin has been known to induce childbirth. In view of the microscale and complexity of the reactions involved, and remembering that there is not even an accurate model for the conformations of these peptides, it is perhaps not surprising that the molecular details of the ways in which they manifest their activities are unknown. However, from the study of synthetic analogues, a great deal is known about the relative importance of the various chemical groups in the parent hormones.

In the oxytocin–vasopressin series, natural peptides occur with residue variations in the 3, 4 and 8 positions. It follows that the specific side chains in these positions are not essential for the compound to possess biological activity and this is confirmed by studies with synthetic compounds. The nature of the residues in the 3 and 8 positions does determine which of the mammalian hormones the analogue most resembles and by modifying these residues, compounds with intermediate properties can be produced (Table 7.1).

Table 7.1. Comparison of hormones in the oxytocin–vasopressin series. Activities (approximate) are expressed in international units per mg. (a) = oxytocin; (b) = oxypressin (only synthetic); (c) = vasotocin (bony fishes); (d) = vasopressin

	Residue 3	8	Oxytocic activity	Milk let-down activity	Rat-pressor activity	Antidiuretic activity
(a)	Ile	Leu	500	500	5	5
(b)	Phe	Leu	20	60	3	30
(c)	Ile	Arg	115	212	245	250
(d)	Phe	Arg	25	100	400	400

Synthetic analogues have been studied in which each of the reactive side chains in turn has been modified to test the importance of the various functional groups. Thus, 1-deaminooxytocin was synthesized to test the significance of the *N*-terminal amino group, and 9-decarboxamido-oxytocin to test the importance of the *C*-terminal carboxamido group. Groups which are not individually necessary for an analogue to possess oxytocin-like activity include the *N*-terminal amino group, the tyrosine phenolic group and the γ-carboxamido group of the glutamine residue; the presence of the carboxamido groups of the asparagine and glycine residues is essential. It should be noted that parts of the molecule which are not essential for biological activity often do contribute in some way to the response. Perhaps they do this by facilitating binding with a receptor molecule, by influencing the conformation of the hormone itself, or in ways not defined. In their absence, biological activity is diminished but not absent. The one exception to this is the terminal amino group. 1-Deaminooxytocin is appreciably more active than oxytocin itself. The reason for this is not known but it may be related to the fact that deamino-oxytocin is not susceptible to aminopeptidase deactivation.

It was once thought that the disulphide bridge played a functional role in the manifestation of oxytocic activity because reduced oxytocin and thioethers derived from it (Exercise 5.6*) are inactive. Recent studies with synthetic analogues of 1-deaminooxytocin, in which either or both of the sulphur atoms are replaced by a methylene group, show conclusively that the presence of the disulphide is not necessary for biological activity. Very recently, diselenooxytocin, in which each of the sulphur atoms is replaced by selenium, has been found to be somewhat more active than oxytocin. Since this compound crystallizes well and has built-in reference atoms, there is a good chance that X-ray crystallographic studies of this compound will be successful. On the basis of the above evidence it may be concluded that the importance of the disulphide group in the oxytocin series is conformational rather than functional.

Insulin also has a twenty atom disulphide ring and in this case too the role of the disulphide bridge is presumably to stabilize a specific conformation. Studies with a synthetic analogue of insulin in which a cystathionine residue **(301)** is present instead of cystine at positions 6 and 11 in the A chain, show that the intrachain disulphide bridge of insulin plays no active role in the biological response.

$$\overset{+}{N}H_3.\underset{\displaystyle CO_2^-}{\underset{|}{CH}}.CH_2.CH_2.S.CH_2.\underset{\displaystyle CO_2^-}{\underset{|}{CH}}.\overset{+}{N}H_3$$

(301)

Other peptide hormones which have been studied also possess some particular residues which are important for biological activity and others which can be changed without loss of activity, but no generalizations are possible about the likely significance of a given residue in a new structure. As opposed to oxytocin, in which the whole nonapeptide structure is essential, large regions of some peptide hormones seem superfluous so far as the biological response is concerned. Thus, a peptide with the *N*-terminal twenty-four amino acid residue sequence of adrenocorticotrophic hormone possesses the same activity as the full thirty-nine residue hormone. Even more strikingly, the *C*-terminal tetrapeptide amide fragment of the seventeen residue hormone, gastrin, displays the same range of pharmacological responses as the parent hormone. The role of the apparently superfluous regions of these molecules is uncertain. Possibly it is related to transport mechanisms or to the protection of the hormone from enzymic inactivation.

It is possible that peptide hormones bring about a biological response by acting as coenzymes, like glutathione, or by combining with specific proteins to form active enzymes, in the way that *S*-peptide activates ribonuclease S. Many compounds are known to modify the properties of vital biological membranes and changes in, for example, membrane permeability and the enzymic properties of membranes have been observed. Possibly the hormones exert their effect at this level. Certainly this is the case with depsipeptide antibiotics. Their antimicrobial activity can be related directly to their ability to induce active metal ion transport through cellular and mitochondrial membranes. They exhibit ion–dipole binding of metal cations with a high degree of specificity. It is also tempting to speculate that the activity of antamanid (**302**), which is able to protect mice from the toxic action of peptides isolated from the death cap toadstool (*Amanita phalloides*), is due to a specific type of adsorption. Antamanid has no side chains which would normally be reckoned chemically reactive, but its structure cannot be haphazardly modified without consequent loss of activity. For example, the cyclopeptide in which the Ala.Phe sequence is reversed, is completely devoid of activity.

→Pro.Phe.Phe.Val.Pro─┐
└Pro.Phe.Phe.Ala.Pro←┘

(**302**)

CHAPTER EIGHT

Biosynthesis

The organic chemist has an ulterior motive in looking at biosynthesis. By studying the way compounds are prepared in nature, he hopes to obtain information which will help him in the laboratory. Thus, a knowledge of biosynthesis might help him to determine the structures of newly isolated compounds and to develop new synthetic or partially-synthetic methods of preparing them.

However, the intrinsic interest of biosynthesis is considerable and the biosynthesis of proteins is particularly fascinating. Furthermore, it is supremely important because it is through the controlled biosynthesis of proteins that the chromosomes exercise their control over the development and daily metabolism of the organism. Protein biosynthesis, especially the production of the appropriate enzyme at the right time, underlies the biosynthesis of all other natural products. The elucidation of the mechanism of protein biosynthesis, although it is by no means complete, is perhaps the most remarkable achievement of modern science.

This chapter is intended to give an outline of protein biosynthesis, but since this is an extremely complex subject, the treatment is of necessity superficial and largely descriptive. Nucleic acids, for example, which play a fundamental role in protein biosynthesis, really merit a text book to themselves.

Structure and biosynthesis of nucleic acids

Two main types of nucleic acid are important in protein biosynthesis. One of these, ribonucleic acid (RNA), gives on hydrolysis a mixture of heterocyclic bases, phosphoric acid and ribose; the other, deoxyribonucleic acid (DNA), gives heterocyclic bases, phosphoric acid and 2-deoxyribose. The heterocyclic bases commonly encountered are the purine bases, adenine (**303**) and guanine (**304**) and the pyrimidine derivatives, cytosine (**305**), uracil (**306**) and thymine (**307**). Uracil is obtained from RNA, thymine from DNA.

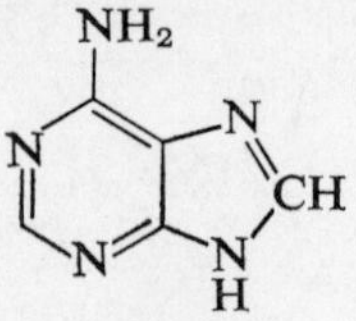

(**303**) 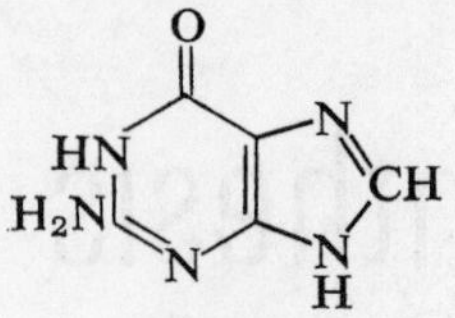 (**304**)

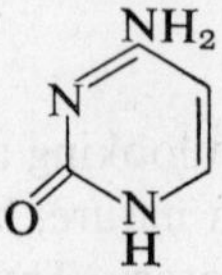

(**305**) 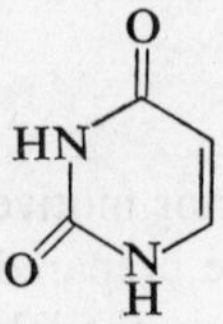 (**306**) 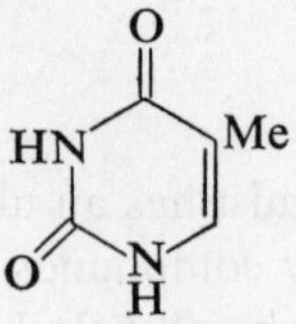 (**307**)

The native nucleic acids are built up of tripartite units called nucleotides. Each nucleotide contains one of the heterocyclic bases condensed to the C_1 atom of a sugar-5-phosphate residue. Adenylic acid (**308**) is an example of a nucleotide. A nucleoside, for example, adenosine (**309**), contains only the base–sugar unit.

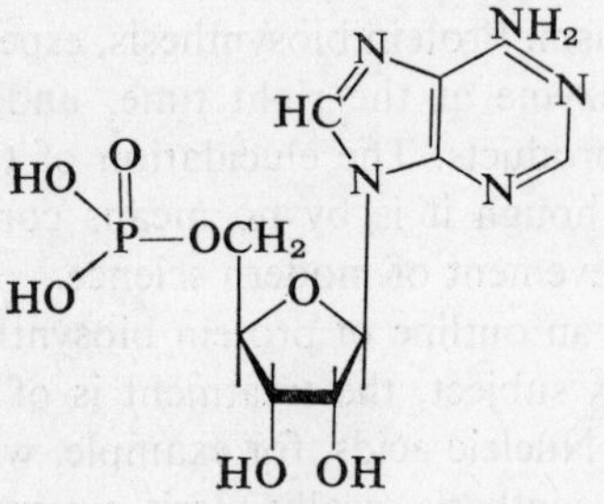

(**308**) 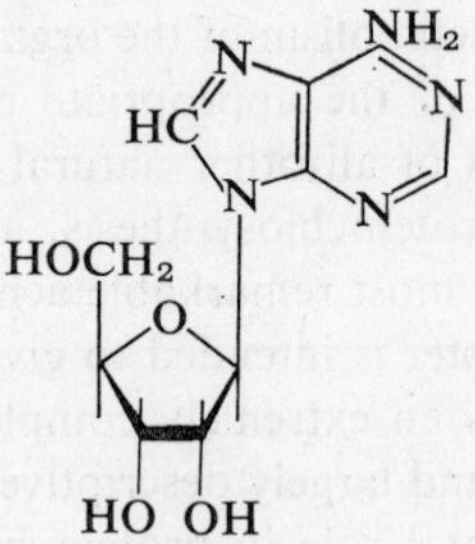 (**309**)

Exercise 8.1

Devise a plausible reaction mechanism for the hydrolysis of adenosine under acidic conditions.

In nucleic acids, the nucleotides are linked by ester bonds which involve the phosphoric acid part of one nucleotide and the 3-hydroxyl group of the ribose in the adjacent nucleotide. They are therefore polynucleotides (**310**). It is interesting that DNA is more stable than RNA to acidic hydrolysis. Presumably, participation of the 2-hydroxyl group in the case of RNA contributes to the cleavage of the 3-*O*-phosphate bond (compare p. 89).

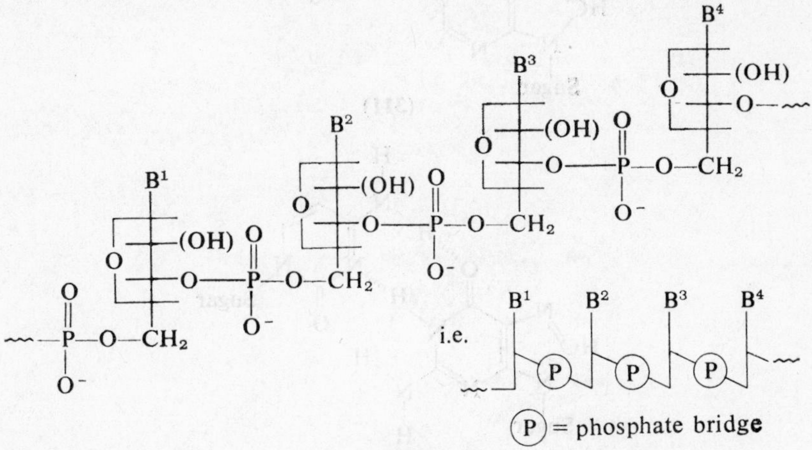

i.e.

Ⓟ = phosphate bridge

(310)

Exercise 8.2

Outline a general scheme of the type given on page 42, to indicate a possible synthetic route to simple nucleotides.

Although the covalent units of the nucleic acids are similar, DNA and RNA differ in molecular weight, in base composition and in secondary and tertiary structure. They also have different functions.

DNA, characteristically found in the nucleus of the cell, is the substance which constitutes the chromosomes. It is a high molecular weight polymer ($M \geqslant 10^6$) which consists of two strands. The strands are held together by hydrogen bonds formed between adenine and thymine (**311**) and between guanine and cytosine (**312**) base pairs. A guanine residue in one strand is always matched by a cytosine residue in the other, an adenine residue in one strand by a thymine residue in the other. Each base pair lies roughly in a plane, perpendicular to the axis of the polymer, whilst the sugar–phosphate chains constitute an intertwined ('double') helix around the stacked base pairs (see Figure 8.1).

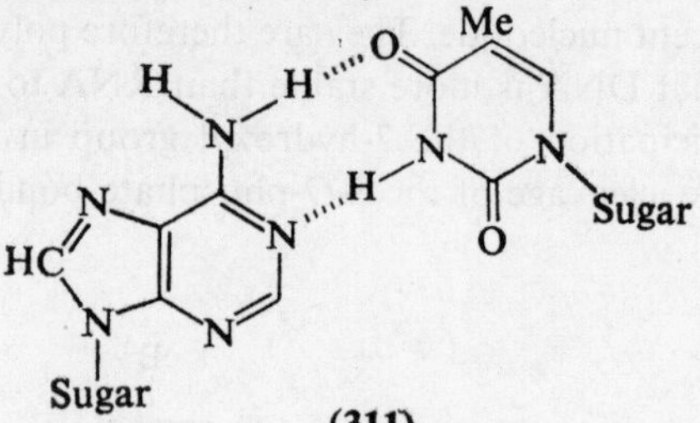

(311)

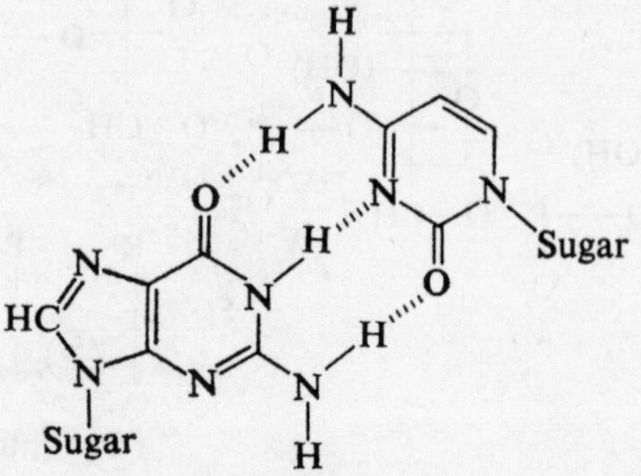

(312)

Since DNA is the genetic material, it must be able to reproduce itself exactly so that each cell in the individual organism will have the same genetic composition, and so that the genetic information can be passed on via the germ cells to the next generation of individuals. The duplication of DNA, which is called *replication*, can be simulated *in vitro* by mixing, under the appropriate conditions, DNA, the four deoxynucleotide triphosphates and an enzyme. This enzyme, DNA-polymerase has been isolated from microorganisms. In the triphosphates, the 5-hydroxyl group is esterified by the phosphoric acid anhydride grouping:

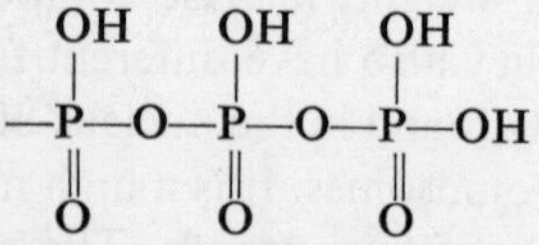

$$-\overset{\displaystyle OH}{\underset{\displaystyle O}{\overset{|}{\underset{\|}{P}}}}-O-\overset{\displaystyle OH}{\underset{\displaystyle O}{\overset{|}{\underset{\|}{P}}}}-O-\overset{\displaystyle OH}{\underset{\displaystyle O}{\overset{|}{\underset{\|}{P}}}}-OH$$

DNA produced in this way possesses the same base sequences (**310**; B^1, B^2, B^3, B^4 etc.) as the DNA molecules with which the system was primed. The introduction of the bases into the growing daughter DNA is, in fact, determined by the hydrogen bonding of the nucleotide triphosphates to the unwinding strands of the template DNA.

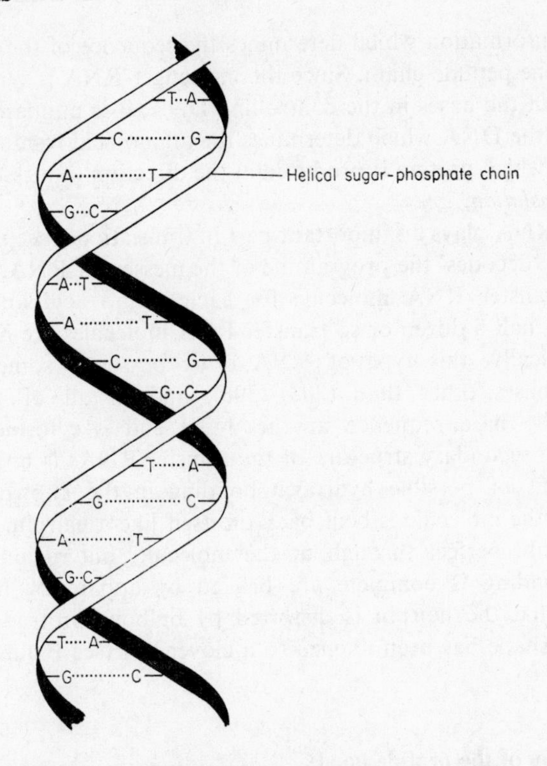

Figure 8.1. DNA double helix (diagrammatic)

DNA therefore acts like a punched-tape programme to control its own replication and the genetic information which it bears is no more than the sequence of its heterocyclic bases. This sequence also programmes the synthesis of proteins, but it does this indirectly, through RNA intermediaries. Three main types of RNA molecules are involved, ribosomal RNA, transfer RNA and messenger RNA. The synthesis of these molecules is programmed by the hydrogen-bonding of ribonucleotide triphosphates to the bases of the DNA, a process known as 'transcription.'

The structural details of ribosomal RNA are imperfectly known. This type of RNA is found in the cell in complex extranuclear components called ribosomes. The ribosomes, which are the actual sites of protein synthesis, are each characteristically composed of two parts, one of which ($M \approx 1 \cdot 7 \times 10^6$) is somewhat larger than the other ($M \approx 0 \cdot 8 \times 10^6$).

Messenger RNA seems to be single-stranded and, perhaps because of this, it is relatively short-lived. The messenger RNA, as its name suggests, carries the information which determines the sequence of the amino acid residues in the peptide chain. Since the messenger RNA is synthesized to the pattern of the bases in the controlling DNA, it is ultimately the base sequence of the DNA which determines the amino acid residue sequence. The step by which protein is produced at the direction of messenger RNA is called *translation*.

Transfer RNA plays an important part in translation since it is transfer RNA which 'decodes' the programme of the messenger RNA. There is at least one transfer RNA molecule for each amino acid and the base sequences of half a dozen or so transfer RNA molecules are now known. Characteristically, this type of RNA often possesses some 'unusual' bases, i.e. bases other than **(303)**–**(306)**, and in all of the known structures the base sequence at the 3-OH end is cytosine, cytosine, adenine. The secondary structure of the transfer RNAs is unknown, but considerations of possible hydrogen bonding in the known sequences suggest that the molecule is bent back on itself like a hairpin. Hydrogen-bonding is not perfect throughout the molecule, but regions in which hydrogen-bonding is complete are broken by apparently non-bonded regions so that the hairpin is distorted by bulbous protruded regions. The overall shape has been likened to a clover leaf (see Figures 8.2a and b).

The formation of the peptide bond

In protein synthesis the amino acids, in the presence of a specific enzyme and adenosine triphosphate, become bound via an ester bond to the 3-hydroxyl group of their own particular transfer RNA (Scheme 8.1). In view of the specificity of this step, an act of recognition must be involved, but, although the nature of this recognition is not yet understood, it need be no more unusual than the specificity of enzymes for particular substrates.

The aminoacyl transfer RNA and one end of the messenger RNA then become bound to the ribosome. In this crucial step, the appropriate aminoacyl transfer RNA is selected by the messenger RNA by a base-pairing mechanism which involves a sequence of three bases in the messenger RNA and three bases in the transfer RNA. Thus, each three consecutive bases in the messenger RNA code for a specific transfer RNA and hence for a particular amino acid.

Once this initial binding has been accomplished, the messenger RNA together with the aminoacyl transfer RNA, 'moves' further into the ribosome and the next three bases of the messenger recognize and bind the

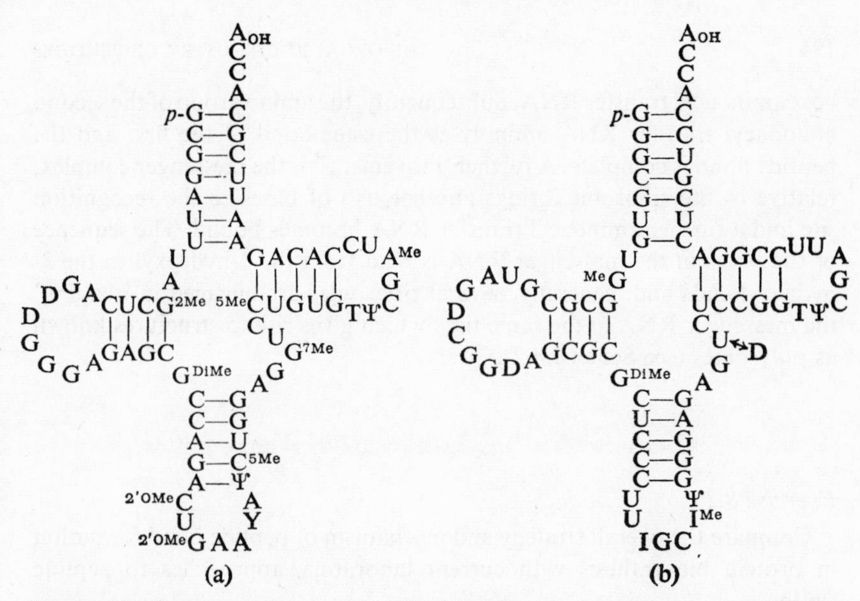

Figure 8.2. Bases in t-RNA. (a) t-RNA^{Phe} (yeast) (b) t-RNA^{Ala} (yeast). A, adenosine; A^{Me}, position of methyl group unknown; Y, fluorescent modified adenosine. G, guanosine; G^{2Me}, 2-methylguanosine; G^{7Me}, 7-methylguanosine; G^{DiMe}, *N*(2)-dimethylguanosine; 2′OMeG, 2′-*O*-guanosine. C, cytidine; C^{5Me}, 5-methylcytidine; 2′OMeC, 2′-*O*-methylcytidine. U, uridine; U*, 4-thiouridine. ψ, pseudouridine; Z, inosine; T, thymidine (After M. J. Waring, *Annual Reports of the Chemical Society, London* (1968) **65**, 555. Reproduced by courtesy of the author and publishers)

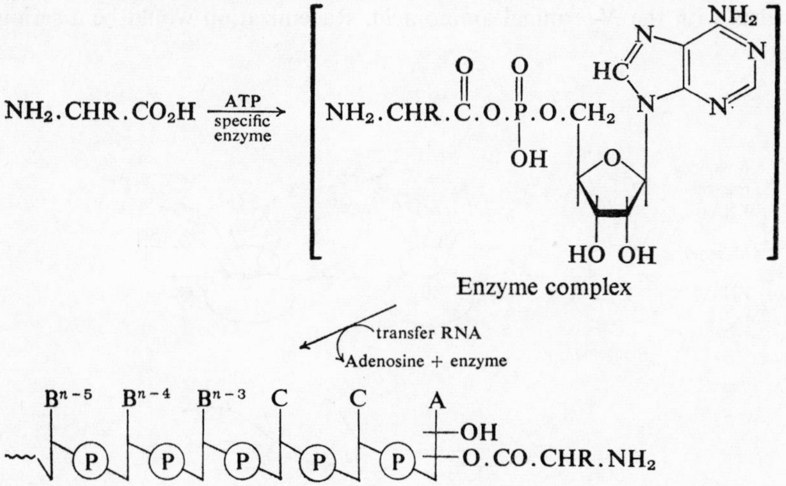

$NH_2.CHR.CO_2H \xrightarrow[\text{specific enzyme}]{\text{ATP}}$ [$NH_2.CHR.C(=O).O.P(=O)(OH).O.CH_2$–adenosine]

Enzyme complex

↙ transfer RNA; Adenosine + enzyme

B^{n-5} B^{n-4} B^{n-3} C C A —OH, —O.CO.CHR.NH_2 (chain: P P P P P)

[C = cytosine; A = adenine]

Scheme 8.1. Formation of the transfer RNA ester of the amino acid next aminoacyl transfer RNA. Subsequently, the amino group of the second aminoacyl transfer RNA aminolyses the ester bond of the first and the peptide bond is complete. A further 'movement' of the messenger complex, relative to the ribosome, brings another trio of bases to the recognition site and a further aminoacyl transfer RNA becomes bound. The sequence of the bases in the messenger RNA is read from the 5-hydroxyl to the 3-hydroxyl ends and, generally, several ribosomes are engaged in 'reading' the messenger RNA at the same time which gives rise to structures known as polysomes (see Scheme 8.2).

---

*Exercise 8.3**

Compare the overall strategy and mechanism of peptide bond formation in protein biosynthesis with current laboratory approaches to peptide synthesis.

The biosynthesis of proteins is therefore a stepwise process in which the peptide chain is built up one amino acid residue at a time. It differs fundamentally from the favoured laboratory strategy in that the synthesis starts with the *N*-terminal amino acid. Racemization would be a serious

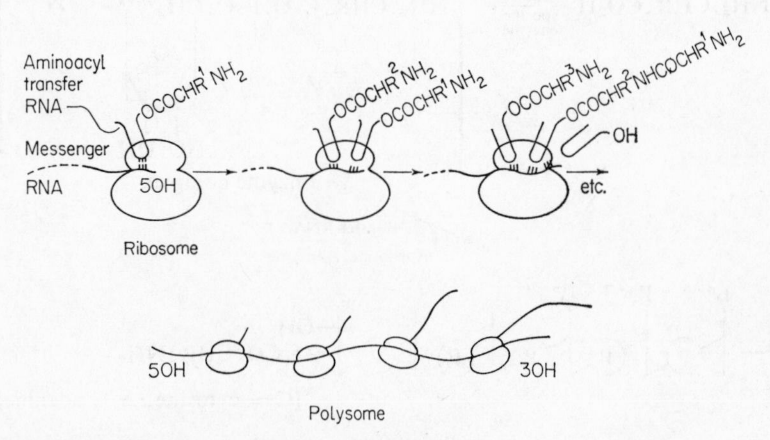

Scheme 8.2. Formation of the peptide bond

disadvantage in a laboratory synthesis which adopted this approach (see p. 101), but it does not occur in biosynthesis. Perhaps the peptide ester is held by the ribosome in a conformation which precludes oxazolone formation. Otherwise, it may be that ribose esters of this type are not activated for oxazolone formation and only undergo aminolysis readily because of anchimeric assistance from the 2-hydroxyl group (compare p. 89).

As described above, protein biosynthesis does not involve the use of protecting groups. Thus, the amino group of the aminoacyl transfer RNA is apparently not blocked. It is not difficult to imagine that steric factors could inhibit the approach of this amino group to the ester carbonyl group, but it is also possible that protecting groups are sometimes involved. Many proteins from microorganisms possess an *N*-terminal methionine residue and an enzyme has been isolated which formylates methionyl transfer RNA. It has been suggested that *N*-formyl methionine is therefore a protecting group. The feasibility of cleaving methionyl peptide bonds selectively, even by chemical methods, has already been discussed (see p. 32).

The genetic code

The sequence of bases in the messenger RNA which recognizes a given aminoacyl transfer RNA is called a *codon*; the sequence of bases in the transfer RNA which is involved in this recognition is the *anticodon* or *nodoc*. Since each codon only contains three bases, whereas there are four different bases in the messenger RNA structure, sixty-four codons, i.e. trinucleotide sequences, are possible. However, only twenty or so amino acids are commonly occurring and it is found that several codons code for the same amino acid. The trinucleotide-amino acid identification key, the *genetic code*, is therefore said to be *degenerate*. This degeneracy, in turn, implies the existence of more than one transfer RNA for each amino acid. The code is summarized in Table 8.1.

It is seen from the code that, for example, the sequence U.U.U. or U.U.C. in messenger RNA will code for phenylalanine. If one of the bases in the sequence is changed, the amino acid inserted in the peptide will be different. Thus A.U.U. or A.U.C. would code for isoleucine. Species variations in proteins are usually of this type, indicating that, in the course of evolution, one of the bases in the polynucleotide has been transposed for another. The evolution of a given peptide can often be deduced if sufficient comparative information is available (Figure 8.3).

Table 8.1. The genetic code. (U = uridine, C = cytosine, A = adenine and G = guanine) (F. H. C. Crick, *Proceedings Royal Society* (1967B) **167**, 334. Reproduced by courtesy of the author and publishers)

5-OH base	Middle base U	C	A	G	3-OH base
	Phe	Ser	Tyr	Cys	U
U	Phe	Ser	Tyr	Cys	C
	Leu	Ser	—	—	A
	Leu	Ser	—	Try	G
	Leu	Pro	His	Arg	U
C	Leu	Pro	His	Arg	C
	Leu	Pro	Gln	Arg	A
	Leu	Pro	Gln	Arg	G
	Ile	Thr	Asn	Ser	U
A	Ile	Thr	Asn	Ser	C
	Ile	Thr	Lys	Arg	A
	Met	Thr	Lys	Arg	G
	Val	Ala	Asp	Gly	U
G	Val	Ala	Asp	Gly	C
	Val	Ala	Glu	Gly	A
	Val	Ala	Glu	Gly	G

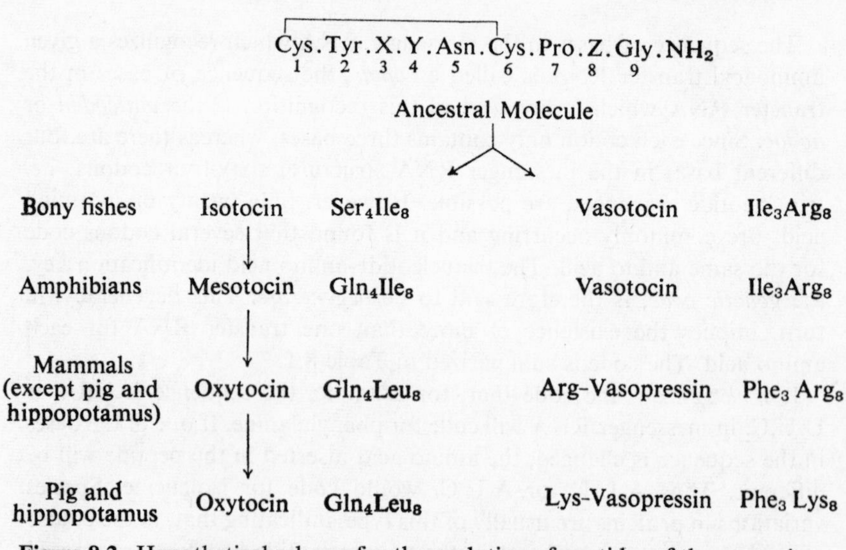

Figure 8.3. Hypothetical scheme for the evolution of peptides of the oxytocin–vasopressin series. (After R. Acher, *Angewandte Chemie*, *Int Ed.* (1966) **5**, 805. Reproduced by courtesy of the author and publishers)

The gaps in the genetic code are real gaps. It seems likely that they represent punctuation in the messenger and are associated with the termination and possibly initiation steps of protein synthesis. Only one codon, A.U.G., codes for methionine in the table although the *N*-terminal formylmethionine mentioned above must presumably be distinguished from methionine which is to be incorporated within the peptide chain. In fact, G.U.G., which codes for valine internally also codes for formylmethionine when it is in a terminal position. However, A.U.G. seems to code for two different methionyl transfer RNAs. One of these can be formylated in the biological system whereas the other cannot. The sequence A.U.G. in a terminal position must be able to recognize the former unequivocally and to reject the latter. It seems probable that it is able to do this with the assistance of the smaller piece of the ribosome. This piece seems to associate first with the messenger RNA and it is probably this fragment in association with the A.U.G. or G.U.G. sequence which recognizes the formylmethionyl transfer RNA. The larger piece of the ribosome seems to join in subsequently.

Generality of the scheme

Although most information about protein biosynthesis has been gleaned from studies with microorganisms, it seems probable that the basic mechanism which has been described operates for the production of proteins in all living organisms. The biosynthesis of smaller peptides is much less understood. Certainly some small peptides, for example, angiotensin (p. 127) and bradykinin (p. 127) are initially part of precursor proteins from which they are subsequently detached, but other small peptides cannot be accounted for in this way. It is impossible to see how the 'unusual' features of many of the antibiotics could be programmed by the mechanism described above. The biosynthesis of antibiotics seems in many cases to compete with protein biosynthesis for the available amino acids and in periods of active protein synthesis, antibiotic production is low.

Little is known of other mechanisms of peptide biosynthesis, but with antibiotics, at least, it seems likely that the primary sequence is not dictated rigidly, but is determined by the amino acids available. Looking at the structures of some of these antibiotics, for example, etamycin and staphylomycin, (p. 140), it is tempting to speculate that the aminoacyl insertion reaction (pp. 39, 135) might be involved in their biosynthesis.

Chemical implications

It will be clear by now that, in the case of proteins, a knowledge of the biosynthetic route does not help in structural elucidation. True, if the sequence of the bases in the messenger RNA were known it would be possible to deduce the primary structure of the protein, but it is more difficult to sequence the polynucleotide than the protein.

Nor does the biosynthetic mechanism suggest immediately a new synthetic approach to proteins. However, biosynthetic processes may well be used in the future for the industrial production of proteins. There is evidence that each of the biosynthetic steps can be carried out in cell-free preparations. It would only be necessary to isolate or synthesize a small amount of the parent DNA and an indefinite further amount could be obtained by the process of *replication*. Subsequently, by harnessing the *transcription* and *translation* machinery, the appropriate protein could be produced on an unprecedented scale. Biosynthesis sets formidable standards. It is estimated that a chain of 200–300 amino acid residues is synthesized in 30 seconds. Reproducibility is virtually one hundred per cent. Expert though he has become, the day is still far removed when the synthetic chemist will be able to make comparable claims for his techniques.

Bibliography

General texts and reviews

1. J. P. Greenstein and M. Winitz, *The Chemistry of the Amino Acids*, Wiley, New York (1960–1).
2. E. Schröder and K. Lubke, *The Peptides*, Academic Press, London (1965–1966) **1, 2.**
3. H. Neurath, *The Proteins*, Academic Press, London, 2nd Edition (1964).
4. H. D. Springall, *The Structural Chemistry of Proteins*, Butterworths, London (1954).
5. E. J. Cohn and J. T. Edsall, *Proteins, Amino Acids and Peptides*, Reinhold, New York (1943–8).
6. *Annual Reports of the Chemical Society*, London: P. M. Hardy (1968, 1969); H. D. Law (1967, 1966, 1965); R. C. Sheppard (1963); D. G. Smyth (1965, 1963). See also *Amino Acids, Peptides and Proteins*, Ed. G. T. Young, Specialist Report of the Chemical Society (1969) **1** (1970) **2.**
7. Proceedings of the Sixth European Peptide Symposium, *Peptides*, Athens, Greece (1963), Ed. L. Zervas, Pergamon Press, Oxford (1966).
8. Proceedings of the Eighth European Peptide Symposium, *Peptides*, Noordwijk, The Netherlands (1966), Ed. H. C. Beyerman, A. van den Linde and W. Maassen van den Brink, North Holland Publishing Co. Amsterdam (1967).
9. Proceedings of the Ninth European Peptide Symposium, Orsay, France (1968); *Peptides, 1968*, Ed. E. Bricas, North Holland Publishing Co., Amsterdam (1968).

Chapter 1

10. Nomenclature: R. Schwyzer, J. Rudinger, E. Wünsch and G. T. Young. *Fifth European Peptide Symposium*, Ed. G. T. Young, Pergamon Press, Oxford (1963).
11. Amino acid analysis: S. Blackburn, *Amino Acid Determination*, Arnold, London (1968).

Chapter 2

12. General: I. Harris and V. M. Ingram in *Analytical Methods of Protein Chemistry*, Ed. P. Alexander and R. J. Block, Pergamon Press, Oxford (1960) **2**, 421.
13. General: E. O. P. Thompson, *Advances in Organic Chemistry*, Ed. R. A. Raphael, E. C. Taylor and H. Wynberg, Wiley–Interscience, London (1960) **1**, 149.
14. General: J. Leggett-Bailey, *Techniques in Protein Chemistry*, Elsevier, 2nd Edition (1967).
15. General: *Methods in Enzymology*, Ed. C. H. W. Hirs, Academic Press, London (1967) **11**.
16. DNP, Edman and carboxypeptidase: H. Fraenkel-Conrat, J. I. Harris and A. L. Levy in *Methods of Biochemical Analysis*, Ed. D. Glick, Wiley–Interscience, London (1955) **2**, 360.
17. Dansylation: W. R. Gray, ref. 15, p. 137.
18. Edman: P. Edman and G. Begg, *European Journal Biochem.* (1967) **1**, 80; R. F. Doolittle, *Biochemical Journal* (1965) **94**, 742; D. G. Smyth and D. F. Elliott, *Analyst* (1964) **89**, 81.
19. Selective Chemical Cleavage: B. Witkop, *Advances in Protein Chemistry* (1961) **16**, 221.
20. *Mass spectrometry of peptides*: E. Lederer and B. C. Das, see ref. 8, p. 131.

Chapters 3 and 4

21. General: M. Bodanszky and M. A. Ondetti, *Peptide Synthesis*, Wiley–Interscience, London (1966).
22. General: M. Goodman and G. W. Kenner, *Advances in Protein Chemistry* (1957) **12**, 465.
23. General: H. N. Rydon, *Peptide Synthesis*, Royal Institute of Chemistry Monograph No. 5 (1962).
24. General: J. Meienhofer, *Chimia* (1962) **16**, 385.
25. General: T. Wieland and H. Determann, *Angewandte Chemie* (1963) **75**, 539.
26. Aminoacyl insertion: M. Brenner, see ref. 8, p. 1.
27. Protecting groups: J. F. W. McOmie, *Advances in Organic Chemistry*, Ed. R. A. Raphael, E. C. Taylor and H. Wynberg, Wiley–Interscience, London (1963) **3**, 191.
28. Protecting groups: R. A. Boissonnas, *Advances in Organic Chemistry*, Ed. R. A. Raphael, E. C. Taylor and H. Wynberg, Wiley–Interscience, London (1963) **3**, 159.
29. Coupling methods: See particularly the general references for these chapters and refs. 1, 2, 6, 7, 8 and 9.
30. Coupling methods: N. F. Albertson, *Organic Reactions*, Ed. A. G. Cope, Wiley, New York (1962) **12**, 157.
31. Racemization, G. T. Young. See ref. 8, p. 55.
32. Solid phase: R. B. Merrifield, *Science* (1965) **150**, 178; *Advances in Enzymology*, **32**, in press, and see particularly refs. 8 and 9; Ribonuclease,

Journal of the American Chemical Society (1969) **91**, 501; J. M. Stewart and J. D. Young, *Sólid-Phase Peptide Synthesis*, W. H. Freeman (1969).
33. *N*-Carboxyanhydrides in polymerization: E. Katchalski and M. Sela, *Advances in Protein Chemistry* (1958) **13**, 243.
34. *N*-Carboxyanhydrides in stepwise synthesis: R. Hirschmann and co-workers, ref. 9, p. 139; *Journal of the American Chemical Society* (1969) **91**, 502–508.

Chapter 5

35. General: See particularly refs. 2, 6, 7, 8 and 9.
36. General: H. D. Law, Polypeptides of Medicinal Interest in *Progress in Medicinal Chemistry*, Ed. G. P. Ellis and G. B. West, Butterworths (1965) **4**, 86.
37. General: S. G. Waley, Naturally Occurring Peptides, in *Advances in Protein Chemistry* (1966) **21**, 2.

Chapter 6

38. General: See particularly refs. 2, 6, 7, 8, 9 and 37.
39. General: R. O. Studer, Polypeptide antibiotics of medicinal interest, in *Progress in Medicinal Chemistry*, Ed. G. P. Ellis and G. B. West (1967) **5**, 1.
40. Synthesis of cyclic peptides: R. Schwyzer, *Ciba Foundation Symposium on Amino Acids and Peptides with Antimetabolic Activity*, Churchill, London (1958), p. 171.
41. Cyclodepsipeptides: D. W. Russell, *Quarterly Reviews* (1966) **20**, 559.
42. Hydroxyacyl insertion: M. M. Shemyakin, V. K. Antonov and A. M. Shkrob, ref. 7, p. 319.

Chapter 7

43. General: See particularly ref. 3.
44. General: L. Stryer, Implications of X-ray crystallographic studies of protein structure, in *Annual Review of Biochemistry* (1968) **37**, 25.
45. Enzyme Structure: *Methods in Enzymology*, Ed. C. H. W. Hirs, Academic Press, London (1967) 11.
46. Myoglobin: J. C. Kendrew. *Scientific American* (1961, Dec.); *Science* (1963) **139**, 1259 (Nobel Lecture).
47. Haemoglobin: M. F. Perutz, *Scientific American* (1964, Nov.); *Science* (1963) **140**, 863 (Nobel Lecture).
48. Lysozyme: D. C. Phillips, *Scientific American* (1966) 78. P. Jollés, *Angewandte Chemie, International Edition* (1969) **8**, 227.
49. M. Goodman and N. S. Choi, Conformational Analysis of Polypeptides, see ref. 9, p. 1.
50. Calculation of Conformation: G. N. Ramachandran and V. Sasisekharan, *Advances in Protein Chemistry* (1968) **23**, 283.

51. Function of Enzymes: D. E. Koshland Jrn. and K. E. Neet. The catalytic regulatory properties of enzymes, in *Annual Review of Biochemistry* (1968) **37**, 359.
52. Structure and Function of Proteolytic Enzymes: D. C. Phillips, D. M. Blow, B. S. Hartley and G. Lowe, *Philosophical Transactions of the Royal Society Ser. B* (1970) **257**, 63–266.

Chapter 8

53. General: J. D. Watson, *Molecular Biology of the Gene*, Benjamin, New York (1965).
54. General: V. M. Ingram, *Biosynthesis of Macromolecules*, Benjamin, New York (1965).
55. General: R. F. Steiner, *The Chemical Foundations of Molecular Biology*, Van Nostrand, Princeton, N.J. (1965).
56. General: R. F. Steiner and H. Edel Voch, *Molecules and Life*, Van Nostrand, Princeton, N.J. (1965).
57. Recent developments in molecular biology: see *Annual Reviews of Biochemistry* and F. Gros and M. Revel, *Données récentes sur la biosynthèse des liaisons peptidiques au cours de la traduction génétique*, ref. 9, p. 118.
58. Nucleic acid structure and function. T. L. V. Ulbricht, *Introduction to Nucleic Acids and Related Products*, Oldbourne Press, London (1965).
59. Genetic Code: F. H. C. Crick, *Proceedings of the Royal Society. Ser. B* (1967) **167**, 331. Chemical versus Biochemical process: R. Schwyzer, Biological and chemical synthesis of polypeptides, ref. 8, p. 197.

APPENDIX ONE

The Historical Development of Peptide Chemistry

This list of dates is only intended to show the time-scale of the development of peptide chemistry. It does not indicate the volume of peptide research being carried out at any one time. Over the last fifteen years, literally thousands of papers have been published in this field. The list gives no intimation of this; it takes no account of the staggering number of peptide hormone analogues which have been synthesized during this time and it mentions only a few of the many natural peptides which have been studied.

YEAR

1820 The first of the common α-amino acids (leucine and glycine) isolated from protein hydrolysates.

1865 First recorded attempts to join α-amino acids to make peptides.

1895 Connection between pituitary gland and blood pressure established (Oliver and Schäfer)—vasopressin is the effective agent.

1901 Polyamide hypothesis of protein structure ennunciated (Fischer).

1902 The azide coupling method (Curtius).

1903 The chloroacyl chloride approach to peptide synthesis (Fischer and Otto). Ethyl esters used for carboxyl protection.

1906 First synthesis and polymerization of α-amino acid *N*-carboxyanhydrides (Leuchs).

1909 Functional relationship between pituitary gland and contractions of uterus demonstrated (Dale)—oxytocin is responsible; other biological activities of oxytocin and vasopressin discovered shortly thereafter.

1915 Tosyl amine protection (Fischer).

1916 Link between pituitary gland and degree of pigmentation of tadpole skin demonstrated (Smith)—substance responsible later called melanocyte stimulating hormone.

1921 Glutathione isolated from yeast (Hopkins).

1922 Preparation of insulin (Banting and Best).

1926 Tosyl protection first used in synthesis of a peptide. Phosphonium iodide–hydriodic acid cleavage (Schoenheimer).
Control of adrenal cortex by pituitary gland established (Smith)—substance responsible later called corticotropin.

1928 Almost complete separation of oxytocin and vasopressin achieved (Kamm and coworkers).

1930 Glutathione: structure proposed.

1931 β-Pleated sheet structure (Astbury and coworkers).

1932 Benzyloxycarbonyl amine protection; removed by catalytic hydrogenolysis (Bergmann and Zervas).
Last of the common protein α-amino acids (glutamine and asparagine) isolated (Chibnall and coworkers).

1935 *S*-Benzyl cysteine first used in peptide synthesis (Sifferd and du Vigneaud).
Cleavage of *N*-benzyloxycarbonyl by sodium in liquid ammonia (Sifferd and du Vigneaud).
Synthesis of glutathione (Harington and Mead).

1937 Cleavage of *N*-tosyl with sodium in liquid ammonia (du Vigneaud and Behrens).

1938 Collagens and k-m-e-f groups of fibrous proteins distinguished (Astbury).

1940 First actinomycin isolated from *Streptomyces* spp (Waksman and Wordruft).

1942 Corticotropin: preparation with corticotropic properties isolated from pig pituitary (Li, Simpson and Evans).

1944 Paper chromatography introduced (Consden, Gordon and Martin).
Isolation of Gramicidin S (Gause and Brazhnikova).
Counter-current distribution introduced (Craig).

1945 2,4-Dinitrofluorobenzene method (Sanger).
Structure of gramicidin S proposed (Consden, Gordon, Martin and Synge).

1947 Enniatins isolated from *Fusarium* spp (Plattner and coworkers).

1948 Enniatins A and B: structures proposed (Plattner and coworkers).
Phthalyl group introduced into peptide synthesis (Kidd and Yang; Sheehan and Frank).

1949 Isolation of oxytocin (Livermore and du Vigneaud).

1950 Phenylisothiocyanate stepwise degradation (Edman).
Mixed anhydride coupling method (Wieland and Sehring).

1951 The α-helix (Pauling and Corey).
Mixed anhydride coupling method (Vaughan and Osato; Boissonnas).
Active ester coupling method (Wieland, Schäfer and Bokelmann).
Ion-exchange amino acid analysis (Moore and Stein).
Isolation of arginine–vasopressin (Turner, Pierce and du Vigneaud).

1952 Cleavage of *N*-benzyloxycarbonyl with hydrogen bromide in acetic acid (Boissonnas and Preitner; Ben Ishai and Berger; Anderson, Blodinger and Welcher).
Facile preparation of amino acid benzyl esters (Miller and Waelsch).

1953 Nitroarginine in peptide synthesis (Hofmann, Rheiner and Peckham).
Structure (du Vigneaud, Ressler, Trippett) and synthesis (du Vigneaud, Ressler, Swan, Roberts and Katsoyannis) of oxytocin.

1954 Structure (du Vigneaud, Lawler and Popenoe) and synthesis (du Vigneaud, Gish and Katsoyannis) of arginine–vasopressin.
Automated amino acid analysis (Spackman, Stein and Moore).

1955 *N,N'*-Dicyclohexylcarbodiimide coupling (Sheehan and Hess).
p-Nitrophenyl ester coupling (Bodanszky; Schwyzer, Iselin and Feurer).
Conclusion of decade of investigations which revealed the total primary structure of beef insulin (Sanger and coworkers).
Isolation of glucagon (Staub, Sinn and Behrens).
Pig corticotropin: structure elucidated (Bell and coworkers; White and Landmann).
Collagen: triple helix structures proposed (Crick and Rich; Cowan and coworkers; Ramachandran and Kartha).

1956 Synthesis of gramicidin S and peptide-doubling on cyclization (Schwyzer and Sieber).
Pig α-melanocyte stimulating hormone: structure elucidated (Lee and Lerner).

1957 Introduction of *t*-butoxycarbonyl amine protection (McKay and Albertson; Anderson and McGregor; Schwyzer, Sieber and Kappeler; Carpino).
Structure of glucagon elucidated (Behrens and coworkers).

1959 *t*-Butyl esters used in peptide synthesis (Roeske; Anderson).
Pig β-melanocyte stimulating hormone: structure elucidated (Harris and Roos; Geschwind, Li and Barnafi).
Actinomycin: structures elucidated (Brockmann and coworkers).

1961 Isoxazolium coupling method (Woodward and Olofson).
Crystal structure myoglobin (Kendrew and coworkers).
Isolation of secretin (Jorpes and Mutt).
Actinomycin C_3: synthesis (Brockmann and coworkers).

1963 Solid-phase peptide synthesis (Merrifield).
'Classical' synthesis of insulin (Zahn and coworkers).
Primary structure of ribonuclease elucidated (Smyth, Stein and Moore).
Enniatins A and B: Structures revised and confirmed by synthesis (Plattner, Vogler, Quitt, Studer and coworkers).
Dansyl method described (Gray and Hartley).
o-Nitrophenylsulphenyl amine protection in peptide synthesis (Zervas, Borovas and Gazis).
α-Melanocyte stimulating hormone synthesized (Schwyzer, Costopanagiotis and Sieber).
β-Melanocyte stimulating hormone synthesised (Schwyzer, Iselin, Kappeler, Riniker, Rittel and Zuber).
Pig corticotropin: total synthesis (Schwyzer and Sieber).
Automated Edman degradation first reported (Edman).

1964 Hog gastrin: isolation (Gregory and Tracy), structural elucidation (Gregory, Hardy, Jones, Kenner and Sheppard) and total synthesis (Kenner and coworkers).
Angolide: isolation and structural elucidation (MacDonald and Shannon).

1965 Mass spectrometer studies: fundamental degradation patterns described.
Secretin: elucidation of structure (Mutt, Magnusson, Jorpes and Dahl).

1966 Solid-phase 'synthesis' of insulin (Marglin and Merrifield).
Other 'classical' syntheses of insulin reported (Yueh-ting and coworkers; Katsoyannis and coworkers; Zahn and coworkers).
Synthesis of *S*-peptide of ribonuclease (Hofmann, Smithers and Finn).

1967 Synthesis of glucagon (Wünsch and coworkers).
Synthesis of secretin (Bodanszky, Ondetti, Levine and Williams).
Use of *N*-carboxyanhydrides in stepwise synthesis (Hirschmann and coworkers).

Crystal structure carboxypeptidase A (Lipscombe and coworkers).
Crystal structure ribonuclease A (Kartha and coworkers; Wyckoff and coworkers).
Crystal structure lysozyme (Blake and coworkers; Phillips).
Crystal structure chymotrypsin (Matthews, Sigler, Henderson and Blow).

1968 2-*p*-Diphenylisopropyloxycarbonyl amine protection (Sieber and Iselin).
Solid-phase synthesis of ferredoxin—fifty five amino acid residues (Bayer, Jung and Hagenmaier).
Isolation, structural elucidation and total synthesis of thyrocalcitonin-thyroid hormone which controls plasma calcium level by inhibiting bone degradation (Several groups involved. Synthesis by Rittel and coworkers; and Guttman and coworkers).
Precursor of Insulin isolated—A and B chains joined by link which possesses thirty-three amino acids (Chance, Ellis and Bromer).

1969 Solid-phase 'synthesis' of ribonuclease (Merrifield and coworkers).
N-Carboxyanhydride synthesis of ribonuclease *S*-protein (Hirschmann and coworkers).

APPENDIX TWO

Notes and Answers to Exercises

Chapter 1

1.4 The *gem*-diol is stabilized because of the electron-withdrawing effect of the neighbouring carbonyl groups and the possibility of intramolecular hydrogen-bonding.

1.5

$$\mathbf{(8)} + NH_2.CHR.CO_2H \longrightarrow C_6H_4{<}^{CO}_{CO}{>}C{=}N.CHR.CO_2H \xrightarrow{-CO_2}$$

$$C_6H_4{<}^{CO}_{CO}{>}CHN{=}CHR \xrightarrow{H_2O} C_6H_4{<}^{CO}_{CO}{>}CH.NH_2 \xrightarrow{(+8)}$$

$$(+RCHO)$$

$$C_6H_4{<}^{CO}_{CO}{>}CH{-}N{=}C{<}^{CO}_{CO}{>}C_6H_4 \longrightarrow \mathbf{(9)}$$

1.6 *S*-alanine; *R*-cystine; three (L, D and *meso*). The Fischer designation L and D is usually adequate in peptide chemistry.

1.7

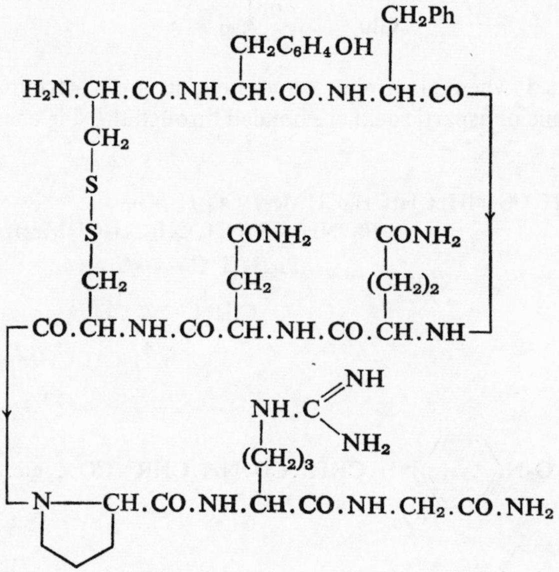

CH_2Ph

$CH_2C_6H_4OH$

$H_2N.CH.CO.NH.CH.CO.NH.CH.CO$–

CH_2

S

S

$CONH_2$ $CONH_2$

CH_2 CH_2 $(CH_2)_2$

–CO.CH.NH.CO.CH.NH.CO.CH.NH–

NH

NH.C

NH_2

$(CH_2)_3$

–N——CH.CO.NH.CH.CO.NH.CH_2.CO.NH_2

Chapter 2

2.8

Hydrazinolysis mixture —(NO_2, NO_2, F-benzene)→ NO_2–C_6H_3(NO_2)–NH.CHR.CO.NH.NH–C_6H_3(O_2N)–NO_2

+ · · · O_2N–C_6H_3(NO_2)–NH.CHR^n.CO_2H

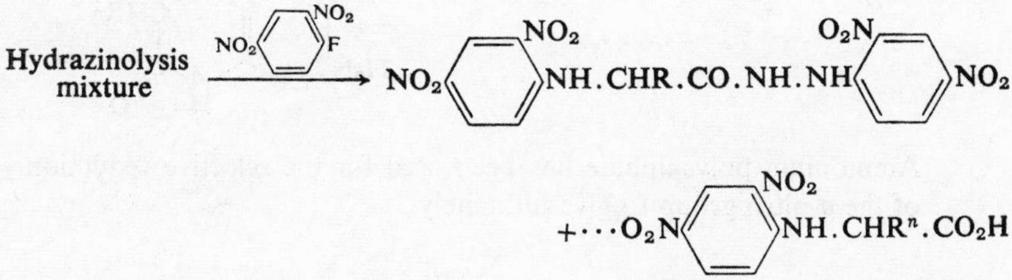

(soluble in aqueous $NaHCO_3$)

Glu.$NHNH_2$ or Asp.$NHNH_2$ obtained when Glu or Asp are in α-linkage in other than the *C*-terminal position.

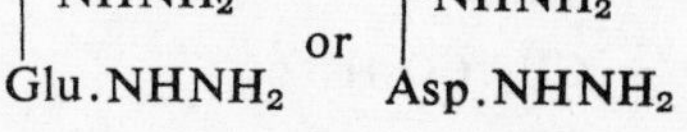

$NHNH_2$ — Glu.$NHNH_2$ or $NHNH_2$ — Asp.$NHNH_2$

obtained when amino acid amide is in other than *C*-terminal position. Glutamine or isoglutamine, and asparagine or isoasparagine not distinguished. Glu or Asp obtained when Glu or Asp are *C*-terminal.

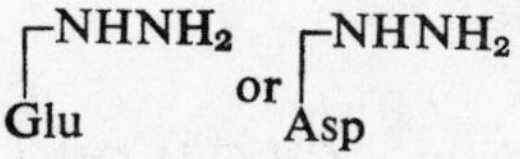

$$\underset{\text{Glu}}{\ulcorner NHNH_2} \quad \text{or} \quad \underset{\text{Asp}}{\ulcorner NHNH_2}$$

obtained when glutamine or asparagine are *C*-terminal or when glutamic or aspartic acid are bonded through the side chain carboxyl.

2.10

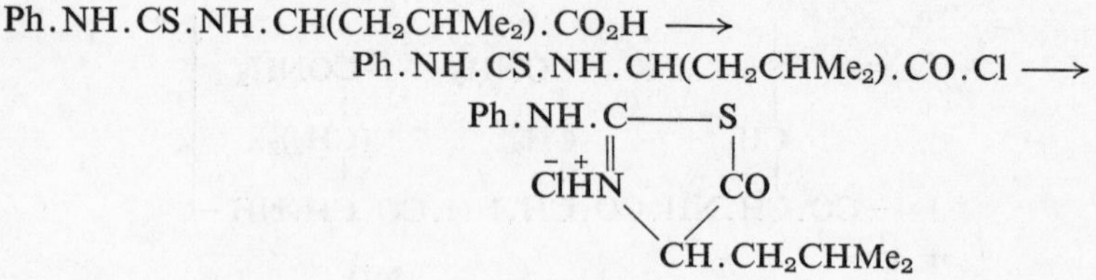

$$Ph.NH.CS.NH.CH(CH_2CHMe_2).CO_2H \longrightarrow Ph.NH.CS.NH.CH(CH_2CHMe_2).CO.Cl \longrightarrow$$

$$\text{Ph.NH.C}\!-\!\!-\text{S} \;;\; \overset{-\;+}{\text{ClHN}}\!=\!\text{C} \;;\; \text{CO} \;;\; \text{ring closed through } CH.CH_2CHMe_2$$

2.11

$$O_2N\text{-}C_6H_3(NO_2)\text{-}NH.CHR^1.CO.NH.CHR^2.CO\ldots\text{etc.} \xrightarrow{H_2/Pt}$$

$$H_2N\text{-}C_6H_3(NH_2)\text{-}NH.CHR^1.CO.NH.CHR^2.CO\ldots\text{etc.} \xrightarrow[\text{(possibly base)}]{\text{heat}}$$

$$H_2N\text{-}C_6H_3\text{<}(NH.CHR^1.C(=O).NH)\text{>}$$

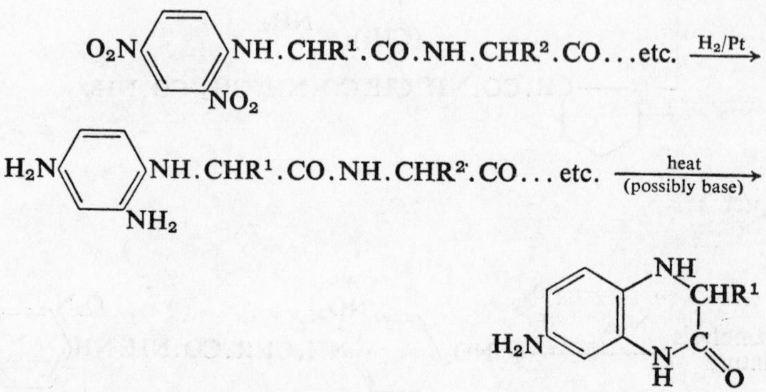

Ammonium polysulphide has been used for the selective reduction of the *o*-nitro group to give ultimately

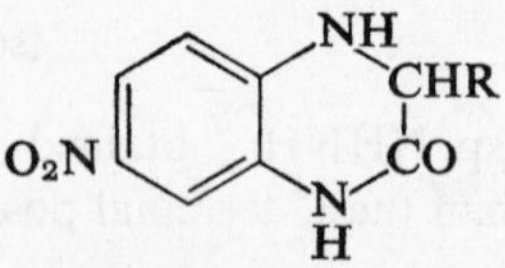

$$O_2N\text{-}C_6H_3\text{<}(NH.CHR.CO.NH)\text{>}$$

2.12

$$R.CO.NH.\underset{|}{\overset{CH_2.CO_2H}{CH}}.CO.NH.CHR^1.CO.R^2 \xrightarrow{CH_2N_2}$$

$$R.CO.NH.\overset{CH_2.CO_2Me}{CH}.CO.NH.CHR^1.CO.R^2 \xrightarrow{\text{base}}$$

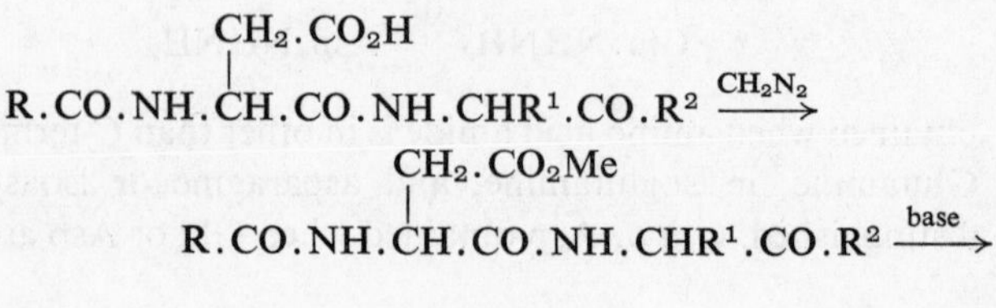

$$\begin{array}{l} \qquad\quad CH_2.CO.NH.CHR^1.CO.R^2 \\ \qquad\quad | \\ R.CO.NH.CH.CO_2^- \end{array} \longrightarrow$$

$$\begin{array}{l} \qquad\quad CH_2.CO.NH.CHR^1.CO.R^2 \\ \qquad\quad | \\ R.CO.NH.CH.CO.N{=}C{=}S \end{array} \xrightarrow{\text{warm}}$$

$$\begin{array}{l} \qquad\qquad CH_2.CO.NH.CHR^1.CO.R^2 \\ \qquad\qquad | \\ R.CO.N\text{——}CH \\ \qquad | \qquad\quad | \\ \quad\; SC \qquad CO \\ \qquad\; \diagdown \quad \diagup \\ \qquad\quad NH \end{array} \xrightarrow{OH^-}$$

$$\begin{array}{l} R.CO_2^- + HN\text{——}CH.CH_2.CO.NH.CHR^1.CO.R^2 \\ \qquad\qquad | \qquad\quad | \\ \qquad\quad\; SC \qquad CO \\ \qquad\qquad\; \diagdown \quad \diagup \\ \qquad\qquad\quad NH \end{array}$$

2.13

$$\begin{array}{l} \qquad\quad CH_2OH \\ \qquad\quad | \\ R.CO.NH.CH.CO.R^1 \end{array} \underset{OH^-}{\overset{H^+}{\rightleftharpoons}} \begin{array}{l} \qquad CH_2.O.CO.R \\ \qquad | \\ \overset{+}{N}H_3.CH.CO.R^1 \end{array} \xrightarrow[pH\approx 5]{\text{2,4-}(NO_2)_2C_6H_3F}$$

$$\text{2,4-}(NO_2)_2C_6H_3\text{-}\begin{array}{l} \quad CH_2.O.CO.R \\ \quad | \\ NH.CH.CO.R^1 \end{array} \xrightarrow{OH^-} \text{2,4-}(NO_2)_2C_6H_3\text{-}\begin{array}{l} \quad CH_2OH \\ \quad | \\ NH.CH.CO.R^1 \end{array} + R.CO_2H$$

2.14

(Structure: 5-bromo spiro-oxindole lactone — Br-substituted indolin-2-one (N–H) spiro-fused at C-3 to a γ-butyrolactone ring, O–C(=O)–CH₂–CH₂.)

2.15 (a) Chain cleavage with *N*-methylpyrrolidone formation is presumably inhibited in the sterically hindered valine situation.

(b)

$C_8H_{13}NO_3$ = $CH_3.CO.NMe.C(CH_2{-}CH_2).CO_2Me$ (cyclopropane ring: C bonded to CH_2—CH_2)

m/e 636 = ⌐MeGlu.MeGly.Pro.MeN.CH(CH₂–[1-methylindol-3-yl]).CO.NMe.C(CH₂—CH₂).CO₂Me

2.16

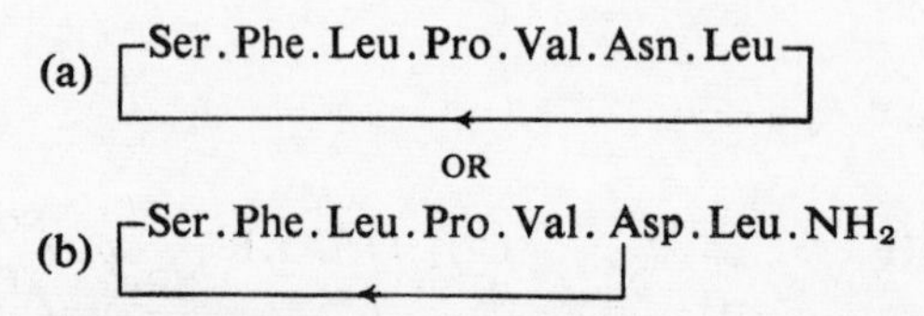

$\alpha \rightarrow \beta$ Aspartyl rearrangement is involved in the degradation. Further evidence showed (a) to be the correct structure.

2.17 First, separation of chains and total sequence determination on each. Enzymic digestion with an endopeptidase or selective chemical cleavage of the intact peptide might then provide fragments containing intact disulphide bonds. Hopefully, these fragments could be related by amino acid analyses or sequence studies to the total sequence.

Chapter 3

3.5 Stability of the carbonium ion is related to charge delocalization. $^+CMe_3$: $(+I)$ effect of methyl groups and, possibly, hyperconjugation—or it may simply be that the methyl groups are less electronegative than hydrogen atoms. $^+CPh_3$: $(+M)$ effect of phenyl rings.

3.8 The reaction is

$$CH_2{=}CMe_2 + H^+ \rightleftharpoons Me_3C^+ \underset{}{\overset{ROH}{\rightleftharpoons}} Me_3C.OR + H^+$$

Presumably the *o*-nitrophenate anion is so deactivated by the $-I$, $-M$ nitro group that alkylation does not occur. Steric hindrance could also be operating.

3.9

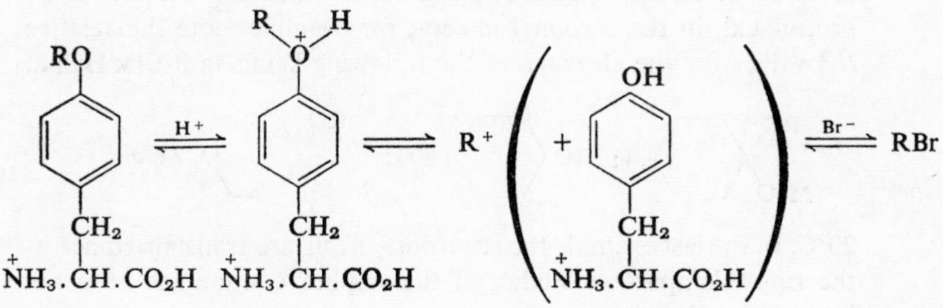

$$p\text{-RO}C_6H_4CH_2\overset{+}{N}H_3.CH.CO_2H \underset{}{\overset{H^+}{\rightleftharpoons}} p\text{-R}\overset{+}{O}(H)C_6H_4CH_2\overset{+}{N}H_3.CH.CO_2H \rightleftharpoons R^+ + \left(p\text{-HO}C_6H_4CH_2\overset{+}{N}H_3.CH.CO_2H \right) \overset{Br^-}{\rightleftharpoons} RBr$$

3.13

$$\bar{C}l.H_3\overset{+}{N}.CH(CH_2.SH).CO_2Me \xrightarrow{\text{dihydropyran}} \bar{C}l.H_3\overset{+}{N}.CH(CH_2\text{—S—}C_5H_9O).CO_2Me \xrightarrow[\text{(ii) } H^+]{\text{(i) } OH^-} NH_2.CH(CH_2\text{—S—}\overset{*}{C}_5H_9O).CO_2H \quad \textbf{(A)}$$

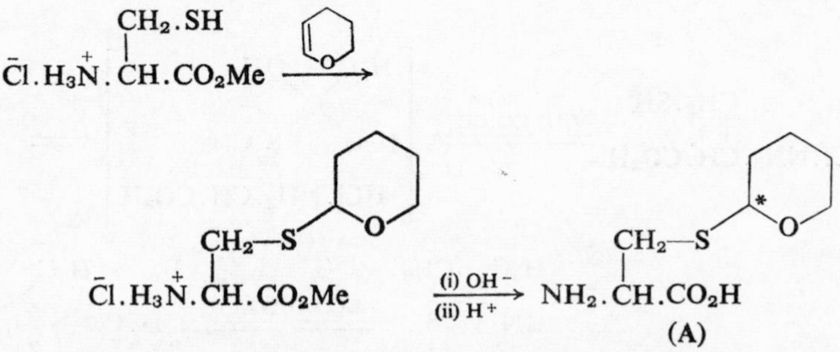

A is a hemithioacetal and hence is acid-labile. Treatment with silver nitrate gives the mercaptide, from which the free thiol can be obtained, for example with H_2S. The carbon indicated in **A** is asymmetric, so that, although **A** is optically active, it is a mixture of diastereoisomers. This optical inhomogeneity could create practical difficulties, for example crystallization problems, if **A** were used in peptide synthesis.

3.14 Hemiacetal and acetal formation are, of course, reversible.

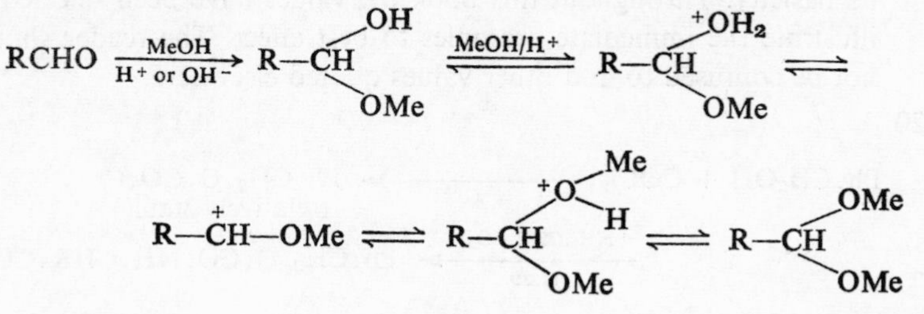

$$RCHO \xrightarrow[H^+ \text{ or } OH^-]{MeOH} R\text{—CH}(OH)(OMe) \overset{MeOH/H^+}{\rightleftharpoons} R\text{—CH}(\overset{+}{O}H_2)(OMe) \rightleftharpoons$$

$$R\text{—}\overset{+}{C}H\text{—OMe} \rightleftharpoons R\text{—CH}(\overset{+}{O}(Me)H)(OMe) \rightleftharpoons R\text{—CH}(OMe)(OMe)$$

Ketals cannot usually be made directly in this way, but, because of the greater nucleophilicity of mercaptans, thioketals are readily formed directly from carbonyl compounds.

These compounds are made more resistant to hydrolysis by the presence of electron-withdrawing groups, which can themselves be protonated, in the carbonyl moiety; for example, note the relative $t^{1/2}$ values for the cleavage of the following ketals in 0·01M HCl at

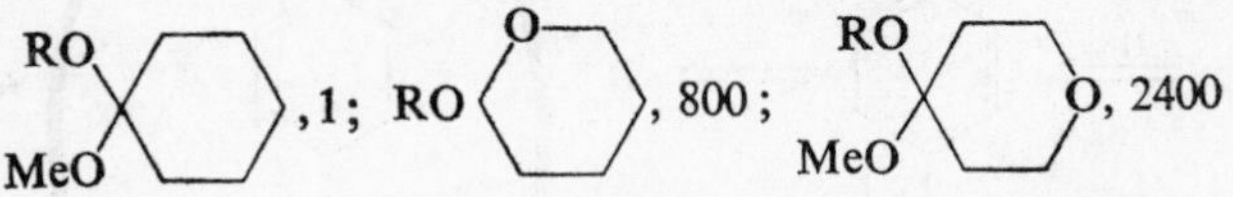

RO, MeO (cyclohexane ketal), 1; RO (tetrahydropyran), 800; RO, MeO (tetrahydropyran-4-one ketal) O, 2400

20°C. In the last example the electronic effects are transmitted across the ring. Comparable studies of the sulphur compounds have not been reported.

3.15

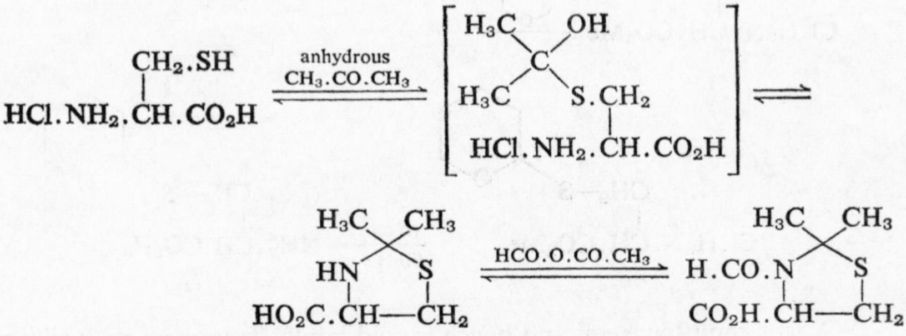

$$HCl.NH_2.CH(CH_2.SH).CO_2H \xrightleftharpoons{\text{anhydrous } CH_3.CO.CH_3} \left[(H_3C)_2C(OH)\text{–}S.CH_2.CH(NH_2.HCl).CO_2H \right] \rightleftharpoons$$

$$(H_3C)_2C\text{(HN–S)}: HO_2C.CH—CH_2 \xrightarrow{HCO.O.CO.CH_3} (H_3C)_2C: H.CO.N—S, CO_2H.CH—CH_2$$

Cleavage by dilute aqueous acid is the reverse process.

3.16 $AH \rightleftharpoons H^+ + B$, $K_a = [H^+][B]/[AH]$. At half neutralization, $[B] = [AH]$ and therefore $pK_a = pH$. The smaller pK_a, the stronger the acid; the larger pK_a, the stronger the base. NOTE: pK values quoted in the literature often show considerable variations, generally due to physical differences at the time of measurement. In Exercise 3.16, dioxan will undoubtedly influence the observed pK values because of its basicity. Throughout this book pK values have been selected to illustrate the immediate examples to best effect. The reader should not be confused to find other values quoted elsewhere.

3.20

$$Ph.CH_2OH + COCl_2 \longrightarrow Ph.CH_2.O.CO.Cl$$

(relatively stable)

$$\xrightarrow[pH \approx 9]{+NH_2.CHR.CO_2^-} Ph.CH_2.O.CO.NH.CHR.CO_2^-$$

3.21 The Hammett equation:

$$\log \frac{k}{k_{\mathrm{H}}} = \rho\sigma \qquad \text{(i)}$$

provides a convenient means for comparing the electronic effects of different substituents in an aromatic species. This equation is derived by comparing rate and equilibrium processes. Thus, if only electronic factors are important, a plot of the logarithms of the dissociation constants of substituted benzoic acids against the logarithms of the rate constants for the saponification reactions of the corresponding ethyl esters is a straight line of the form:

$$\log k = \rho \log K + \mathrm{c} \qquad \text{(ii)}$$

In this equation, k is the rate constant and K the dissociation constant. For benzoic acid and ethyl benzoate:

$$\log k_{\mathrm{H}} = \rho \log K_{\mathrm{H}} + \mathrm{c} \qquad \text{(iii)}$$

By subtraction:

$$\log \frac{k}{k_{\mathrm{H}}} = \rho \log \frac{K}{K_{\mathrm{H}}} \qquad \text{(iv)}$$

For convenience, processes are usually referred to the dissociation of benzoic acid derivatives in water at 25°C. Log K/K_{H}, which is a measure of the electronic effects of the substituents, is termed the *substituent constant* and written σ. Equation (iv) may therefore be rewritten in the form of equation (i). Electron-withdrawing substituents (relative to hydrogen) will have positive σ values, whilst electron-donating substituents will have negative σ values. The magnitude of these values will be a direct measure of the strength of the electronic effect. The factor ρ is called the *reaction constant*. It is a measure of the susceptibility of the reaction to electronic factors.

3.22

Me_3.C.O.CO $<$ CH_2=CH.CMe_2.O.CO $<$

Ph.Me_2C.O.CO $<$ C_6H_5–C_6H_4–CMe_2.O.CO

3.23

Phthalimide potassium salt (N^-K^+) + EtO.CO.Cl ⟶ phthalimide N.CO.OEt $\xrightarrow{+NH_2R}$

HO, NH.R, C, N.C(=O).OEt, C(=O) ⟶ CO.NH.R / CO.NH.C(=O).OEt ⟶

CO, NR, C(HO)(NH.CO.OEt) ⟶ phthalimide N.R

Phthalimide and an α-halocarboxylic acid would give a racemic product.

3.24

Histidine: imidazole (N, NH) $CH_2.CH(NH_2)CO_2H$ $\xrightarrow[\text{(ii) Ph.CH}_2\text{.Cl}]{\text{(i) Na/liquid NH}_3}$

N-benzylimidazole $CH_2.CH(NH_2)CO_2^-$ (N–CH_2.Ph) $\xrightarrow[\text{pH} \approx 9]{\text{Ph.CH}_2\text{.O.CO.Cl}}$ $CH_2.CH(CO_2^-).NH.CO.O.CH_2.Ph$ (imidazole N–CH_2.Ph)

3.25

$H_2\overset{+}{N}$=C(NH_2)–NH–$(CH_2)_3$–$NH_2.CH.CO_2^-$ $\xrightarrow[\text{Ph.CH}_2\text{.O.CO.Cl}]{\text{pH 8–9}}$ $H_2\overset{+}{N}$=C(NH_2)–NH–$(CH_2)_3$–$Ph.CH_2.O.CO.NH.CH.CO_2^-$

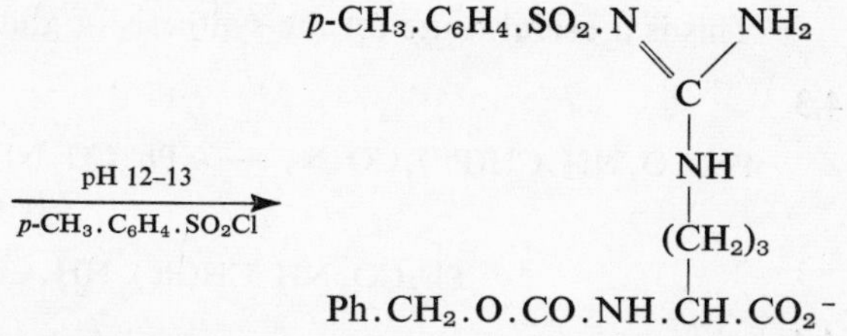

$\xrightarrow[p\text{-}CH_3.C_6H_4.SO_2Cl]{pH\ 12\text{–}13}$ $p\text{-}CH_3.C_6H_4.SO_2.N{=}C(NH_2).NH.(CH_2)_3.CH(NH.CO.O.CH_2.Ph).CO_2^-$

3.26 Spectroscopic data shows that the nitroimino moiety is still present in A; i.e. 1695 cm^{-1}, C=N; 1600 cm^{-1}, NO_2; λ_{max} = 270 nm as in parent compound. This is confirmed by hydrogenolysis which is equivalent to removal of NO_2 group. Hydrolysis to ornithine suggests that the carbon skeleton is also still intact. The changes can be rationalized in the following way:

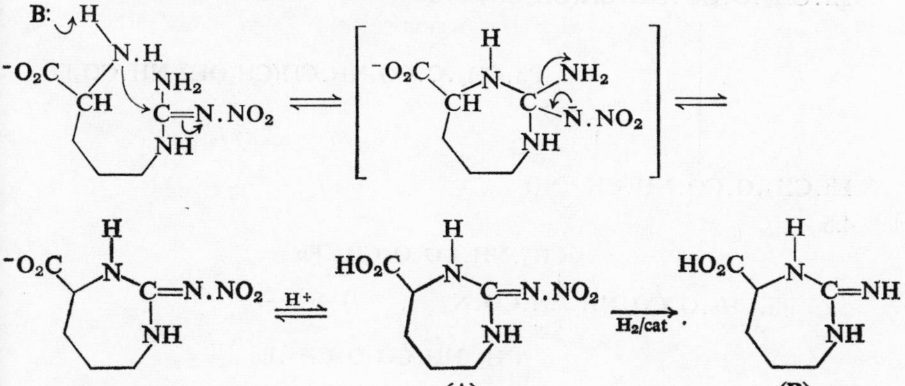

B: H N.H ^-O_2C CH NH_2 C=N.NO_2 NH ⇌ [H ^-O_2C CH N C NH_2 $\bar{N}$.NO_2 NH] ⇌

H ^-O_2C N C=N.NO_2 NH $\underset{}{\overset{H^+}{\rightleftharpoons}}$ H HO_2C N C=N.NO_2 NH (A) $\xrightarrow{H_2/cat}$ H HO_2C N C=NH NH (B)

Chapter 4

4.1

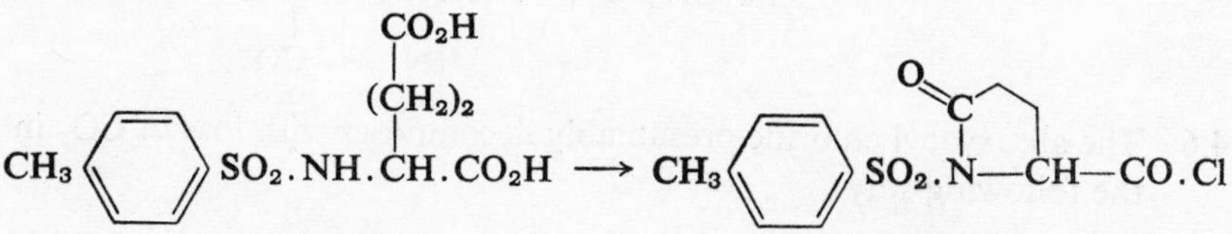

$CH_3C_6H_4SO_2.NH.CH(CO_2H).(CH_2)_2.CO_2H \longrightarrow CH_3C_6H_4SO_2.N$—CH—CO.Cl (A) (cyclic, with O=C—$(CH_2)_2$ ring)

(A) $\xrightarrow{Gly.OEt}$ $CH_3C_6H_4SO_2.N$—CH.CO.Gly.OEt (cyclic, with O=C—$(CH_2)_2$ ring) $\xrightarrow{NH_3}$

Tos.Gln.Gly.OEt ⟶ Gln.Gly

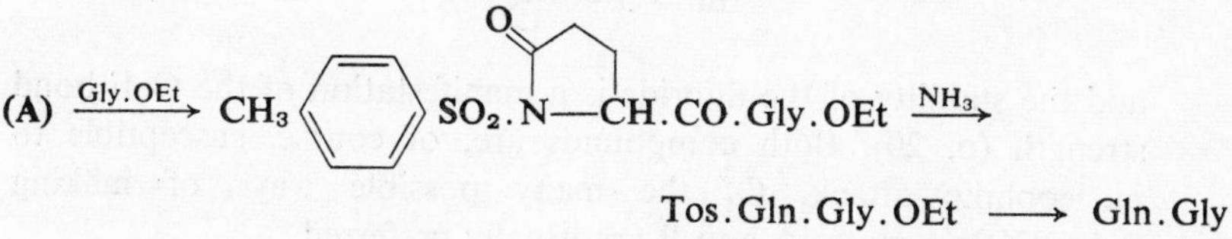

This is a useful route for the synthesis of glutaminyl peptides.

4.3

$$\text{Ph.CO.NH.CH(Pr}^i\text{).CO.N}_3 \longrightarrow \text{Ph.CO.NH.CH(Pr}^i\text{).NCO} \longrightarrow$$
$$(\nu = 2220\ \text{cm}^{-1})$$

$$\text{Ph.CO.NH.CH(Pr}^i\text{).NH.CO.NH.CH(Pr}^i\text{).CO}_2\text{Me}$$

4.4

$\text{Ph.CH}_2\text{.O.CO.NH.CH(CH}_2\text{OH).CO.N}_3$

↘ RNH_2

$\text{Ph.CH}_2\text{.O.CO.NH.CH(CH}_2\text{OH).CO.NH.R} + \text{N}_3\text{H}$

↓

$\text{Ph.CH}_2\text{.O.CO.NH.CH(CH}_2\text{OH).NCO}$

↘ EtOH

$\text{Ph.CH}_2\text{.O.CO.NH.CH(CH}_2\text{OH).NH.CO}_2\text{Et}$

↓

$\text{Ph.CH}_2\text{.O.CO.NH.CH—NH}$ (ring: CH—CH_2—O—CO—NH)

4.5

$$\text{Ph.CH}_2\text{.O.CO.NH.CH(CH}_2\text{.NH.CO.O.CH}_2\text{.Ph).CO.N}_3 \longrightarrow$$

$$\text{Ph.CH}_2\text{.O.CO.NH.CH(CH}_2\text{.NH.CO.O.CH}_2\text{.Ph).NCO} \longrightarrow$$

$\text{Ph.CH}_2\text{.O.CO.NH.CH}$ — CH_2 — $\text{N.CO.O.CH}_2\text{.Ph}$ (ring: CH—CH_2—N—CO—HN—CH)

4.6 The alkoxyacyl chloride presumably decomposes with loss of CO_2 in the following way:

$$\text{Bu}^t\text{—O—C(=O)—Cl}$$

and the stability of the fluoride is a manifestation of the C–F bond strength (p. 20). Both compounds are, of course, susceptible to nucleophilic attack. Of the many possible ways of making $\text{Bu}^t\text{.OCON}_3$, routes A and B are usually preferred.

$$\mathrm{Bu^tOH} \xrightarrow[\mathrm{A}]{\mathrm{COCl_2}} \mathrm{Bu^t.O.CO.Cl} \xrightarrow{\mathrm{NH_2NH_2}} \mathrm{Bu^t.O.CO.NH.NH_2}$$

(unstable)

$$\mathrm{Bu^t.O.CO.NH.NH_2} \longrightarrow \mathrm{Bu^t.O.CO.N_3}$$

$$\mathrm{Bu^tOH} \xrightarrow[\text{(N,N'-carbonyldiimidazole: imidazolyl-CO-imidazolyl)}]{\mathrm{B}} \mathrm{Bu^t.O.\underset{\|}{\overset{}{C}}.N}\text{(imidazolyl)} \ (\mathrm{C{=}O}) \xrightarrow{\mathrm{NH_2NH_2}} \mathrm{Bu^t.O.CO.NH.NH_2}$$

4.7

$$\mathrm{Ph.CH_2.O.CO.NH.CMe_2.CO.NH.CMe_2.CO_2H} \longrightarrow$$

$$\mathrm{Ph.CH_2.O.CO.NH.CMe_2.C{=}N{-}CMe_2{-}CO{-}O}\ \text{(ring: C–O–CO–CMe}_2\text{–N=C)} \xrightarrow{\mathrm{NH_2.CMe_2.CO_2Me}}$$

$$\mathrm{Ph.CH_2.O.CO.NH.CMe_2.CO.NH.CMe_2.CO.NH.CMe_2.CO_2Me}$$

NOTE: No problems of racemization are involved with this amino acid; otherwise this route would not be satisfactory (see p. 97).

4.8 These side reactions arise because of the acidity of the H in sulphonamides (see p. 22 and Exercise 4.1) and from attack at the 'wrong' carbonyl or from failure of the anhydride to form (i.e. $ClCO_2Bu^{sec}$ still present). Other products: *N*-tosyl(*N*-tosylglycyl)-glycine anilide; *N*-*sec*-butoxycarbonyl-*N*-tosylglycine anilide; *sec*-butylcarbanilate; aniline HCl. NOTE: When *N*-ethylpiperidine is used as the base in the formation of the mixed anhydride, tosylglycine (60 per cent) is recovered unchanged and 93 per cent *sec*-butylcarbanilate is formed. Mixed anhydrides of tosylamino acids and pivalic acid form and couple well.

4.12 Lossen rearrangement of the hydroxamic ester occurs:

$$\mathrm{Ph.CH_2.O.CO.Pro.O.N(CO.CMe_3)(H)} \rightleftharpoons$$

$$\left[\mathrm{Ph.CH_2.O.CO.Pro.O.N^-.\overset{O}{\overset{\|}{C}}.CMe_3} \leftrightarrow \mathrm{Ph.CH_2.O.CO.Pro.O.N{=}\overset{O-}{\overset{|}{C}}.CMe_3}\right] \xrightarrow{\mathrm{H^+}}$$

$$\mathrm{(Ph.CH_2.O.CO.Pro)} + \left[\mathrm{Me_3C.\overset{O}{\overset{\|}{C}}{-}\ddot{N}}\right] \longrightarrow$$

$$\mathrm{Me_3C.N{=}C{=}O} \xrightarrow{\mathrm{R.NH_2}} \mathrm{Me_3C.NH.CO.NH.R}$$

Probably rearrangement is predominant in this case because the proline carbonyl is relatively hindered.

4.19 (a) Acidolytic cleavage will not usually be suitable for the synthesis of acid-labile peptides, for example those which contain tryptophan; (b) the incorporation of asparagine and glutamine residues might give rise to ω-nitrile impurities. The use of active ester couplings is therefore better with these particular amino acids; (c) some quarternization of triethylamine by the chloromethylated polymer occurs; (d) some of the amino groups of the individual *C*-terminal amino acid moieties attached to the resin might be more accessible than others. It has been suggested that the coupling of the $(n - 1)$ residue should be reduced in time and unreacted (n) residue blocked by acetylation to reduce the kinetic inhomogeneity of subsequent stages; (e) cleavage of the peptide from the resin requires relatively vigorous conditions. More labile links have been devised; (f) It might prove difficult to isolate the desired peptide. The statistical distribution of the products of such a synthesis can be calculated by using, once again, the binominal expression $(p + q)^n$. For example, if fifty-four coupling steps are carried out with an average yield of 90 per cent for each step, less than one per cent of the required pentapentaconta peptide would be obtained and it would be contaminated with 99 per cent of 'other' peptides; a yield of 99 per cent at each step would give an overall yield of 56 per cent of the required peptide. Fortunately, high yields ($\geqslant$99 per cent) can usually be obtained at each of the individual steps.

4.23

p-$NO_2.C_6H_4.CH_2.O.CO.NH.CH(CO_2H).CH_2$–(imidazolyl: N, CH, NH) ⟶

p-$NO_2.C_6H_4.CH_2.O.CO.NH.CH$–$CH_2$–(imidazole ring, N–CO: OC—N—CH)

(A) (activated amide)

$\xrightarrow{Ph.CH_2NH_2}$

p-$NO_2.C_6H_4.CH_2.O.CO.NH.CH(CO.NH.CH_2Ph).CH_2$–(imidazolyl: N, CH, NH)

4.24

$$\mathrm{Ph.CH_2.O.CO.NH.CH(CH_2.CO.NH_2).CO_2H} \longrightarrow$$

$$\mathrm{Ph.CH_2.O.CO.NH.CH(CH_2{-}C(=O)\ddot{N}H_2).C(=O){-}{}^{18}O.C(=NC_6H_{11})NH.C_6H_{11}} \longrightarrow$$

$$\mathrm{Ph.CH_2.O.CO.NH{-}CH{-}C(=O){-}O{-}C(=NH){-}CH_2\ (ring)} + \mathrm{H^{18}O.C(NH.C_6H_{11})_2} \longrightarrow$$

$$\mathrm{Ph.CH_2.O.CO.NH.CH(CH_2.C{\equiv}N).CO_2H}$$

4.25

$$\mathrm{Ph_3C.NH.CH(CH_2OH).CO_2H} \longrightarrow$$

$$\mathrm{Ph_3C.NH.CH{-}C(=O){-}O{-}CH_2\ (ring)} \xrightarrow{\mathrm{NH_2CH_2Ph}}$$

(reactive β-lactone)

$$\mathrm{Ph_3C.NH.CH(CH_2OH).CO.NH.CH_2.Ph}$$

4.26

$$\mathrm{Ph.CH_2.O.CO.NH.CH((CH_2)_3NH.C(=N.NO_2)NH_2).CO_2H} \longrightarrow$$

$$\text{3-(NH.CO.O.CH}_2\text{.Ph)-1-(C(=N.NO}_2\text{)NH}_2\text{)-piperidin-2-one} \xrightarrow{\mathrm{Pro.OBu^t}}$$

$$\mathrm{NO_2.N{=}C(NH_2){-}N}\text{(pyrrolidine, 2-}\mathrm{Bu^tO_2C}\text{)} + \text{3-(NH.CO.O.CH}_2\text{.Ph)-piperidin-2-one (NH, O)}$$

4.27

$$H_2N{-}C({=}N.NO_2){-}NH_2 \xrightarrow{NH_2NH_2} H_2N{-}HN{-}C({=}N.NO_2){-}NH_2 \xrightarrow{CH_3COCH_2COCH_3} \text{3,5-dimethylpyrazol-1-yl}{-}C({=}N.NO_2){-}NH_2 \xrightarrow{R\bar{N}H}$$

$$R.NH{-}C(\text{3,5-dimethylpyrazol-1-yl})(NH_2){-}\bar{N}.NO_2 \longrightarrow R.NH.C({=}N.NO_2){-}NH_2$$

The pyrazole anion makes a good leaving group.

Chapter 5

5.3 Yes. Glutathione has been prepared successfully in this way. The *N*-trityl group was cleaved selectively at (a) by 5M-hydrochloric acid in acetone (difference between *N*-trityl and *S*-trityl). Since the α-carboxyl group of Tri.Glu is sterically hindered, coupling of the β-carboxyl is favoured. At (b) saponification followed by saturated HCl–chloroform gave glutathione which was isolated as its hydrochloride by ion exchange chromatography.

5.11 The approach, as shown in the diagram on page 223, represents a substantial improvement on the *S*-benzylcysteine approach. The polymer is cleaved in relatively high yield by sulphitolysis and no protecting groups are present which necessitate treatment with sodium in liquid ammonia.

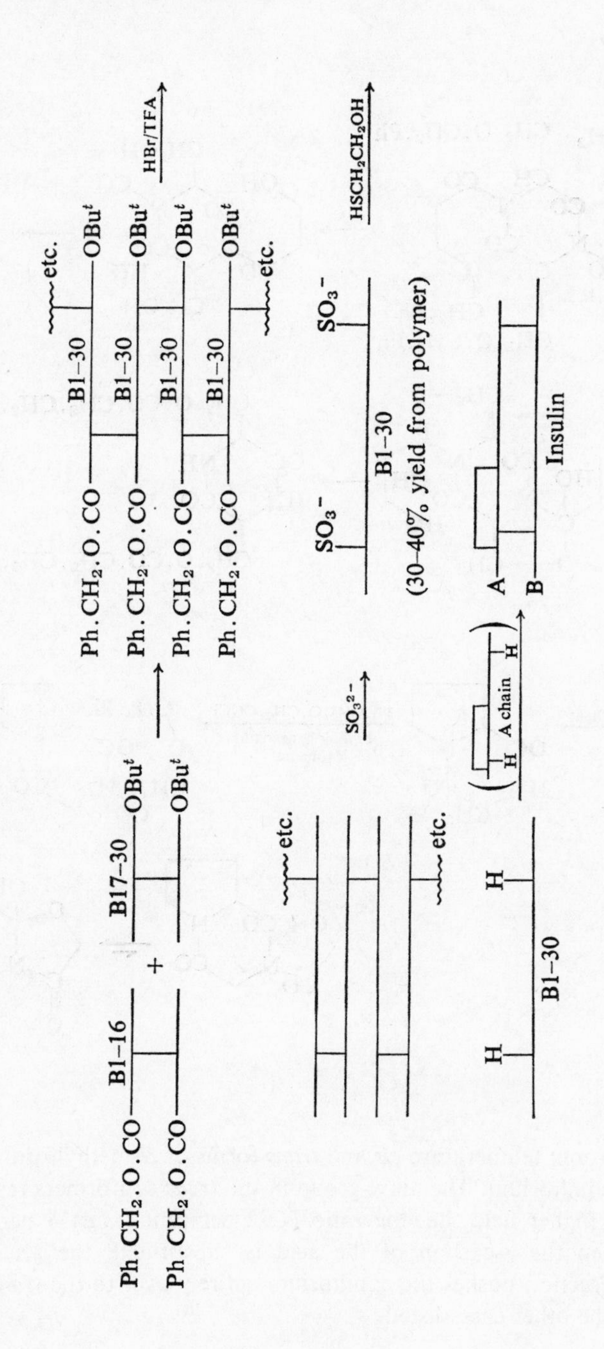
Ph.CH2.O.CO—B1–16
Ph.CH2.O.CO
+
B17–30—OBut
OBut
etc.
Ph.CH2.O.CO—B1–30—OBut
etc.
HBr/TFA
SO3 2−
SO3−
B1–30
SO3−
HSCH2CH2OH
(30–40% yield from polymer)
H
B1–30
H
H
A chain
H
A
B
Insulin

Chapter 6

6.15

Ph.CH$_2$ CH$_2$.O.CH$_2$.Ph
O CH CO
CO N
N CO
CO CH O
CH$_2$.Ph
CH$_2$.O.CH$_2$.Ph
→
CH$_2$OH
OH CO
CO N
N CO
CO HO
CH$_2$OH
→

CH$_2$—O
OH C
CO N
HO OH
N CO
C HO
O—CH$_2$
→
CH$_2$.O.CO.CH$_2$.CH$_2$.OH
OC NH
HN CO
CH$_2$.O.CO.CH$_2$.CH$_2$.OH

6.16

Pro.Gly $\xrightarrow{\text{heat}}$
OC N
HN CO
CH$_2$
$\xrightarrow[\text{boiling benzene or toluene}]{\text{Ph.CH}_2\text{.O.CH}_2\text{.COCl}}$
CH$_2$.Ph
O OC N
CH$_2$ N CO
CO
$\xrightarrow{\text{H}_2\text{/PdC}}$

OH CO N
N CO
CO
⇌
OH
O N
C N CO
O

Chapter 7

7.2 At room temperature *cis* and *trans* forms of *N*-methylformamide are in equilibrium. The alkyl group in the *trans* conformers resonates at the higher field, i.e. the ratio is 92 per cent *trans*: 8 per cent *cis*. When the α-carbon of the acid is substituted, the greater steric interaction pushes the equilibrium entirely over to the *trans* form as in the other cases listed.

7.4 Bifunctional catalysis is involved in the following way:

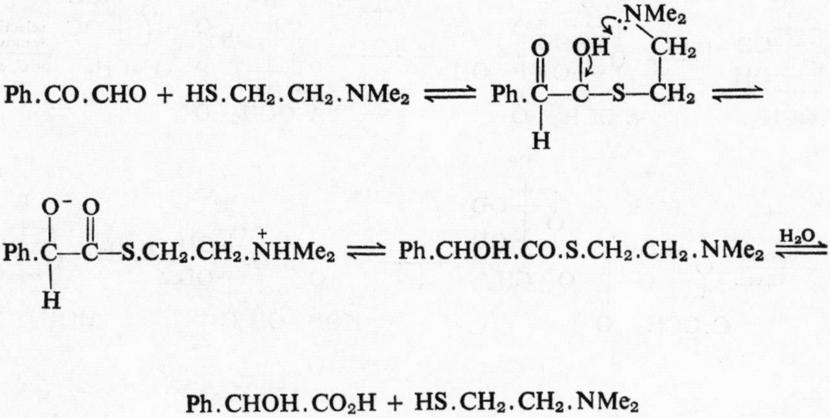

$$Ph.CO.CHO + HS.CH_2.CH_2.NMe_2 \rightleftharpoons Ph.\overset{O}{\overset{\|}{C}}-\underset{H}{\underset{|}{\overset{OH}{\overset{|}{C}}}}-S-CH_2(CH_2NMe_2) \rightleftharpoons$$

$$Ph.\underset{H}{\underset{|}{\overset{O^-}{\overset{|}{C}}}}-\overset{O}{\overset{\|}{C}}-S.CH_2.CH_2.\overset{+}{N}HMe_2 \rightleftharpoons Ph.CHOH.CO.S.CH_2.CH_2.NMe_2 \xrightarrow{H_2O}$$

$$Ph.CHOH.CO_2H + HS.CH_2.CH_2.NMe_2$$

Chapter 8

8.1

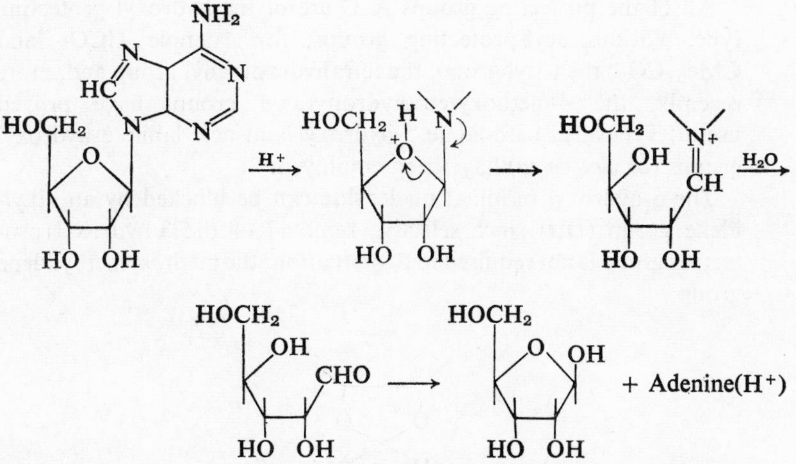

Adenosine (HOCH₂, HO, OH; adenine: NH₂, N, HC, N) $\xrightarrow{H^+}$ protonated intermediate (HOCH₂, H, $\overset{+}{O}$, N; HO OH) $\longrightarrow$ iminium intermediate (HOCH₂, OH, N+, CH; HO OH) $\xrightarrow{H_2O}$

open-chain sugar (HOCH₂, OH, CHO; HO OH) $\longrightarrow$ ribose (HOCH₂, O, OH; HO OH) + Adenine(H^+)

8.2 The building blocks in the synthesis of oligonucleotides are the individual nucleosides. In a sense, these are the counterparts of the amino acids used in peptide synthesis and nucleotide synthesis, like peptide synthesis, is an exercise in the selection of suitable protecting groups and coupling methods. It is not critical whether the synthesis starts from the 3-hydroxyl or from the 5-hydroxyl end since the danger of racemization is not as prominent as in peptide synthesis. Thus, two general routes may be envisaged:

I

B^1 / O / —OB / —OH / A.OCH$_2$ $\xrightarrow{H_3PO_4}$ B^1 / O / —OB / —O—P(=O)(O)—OH / A.OCH$_2$ $\xrightarrow{\text{HOH}_2\text{C-nucleoside}(B^2, \text{OB}, \text{OC})}$ B^1 / O / —OB / —O—P(=O)(O)—O—CH$_2$ / A.OCH$_2$; B^2 / O / —OB / —OC $\xrightarrow{\text{selective removal of C, etc.}}$

II

B^n / O / —OD / —OD / HOH$_2$C $\xrightarrow{H_3PO_4}$ B^n / O / —OD / —OD / HOP(=O)(O)—OH$_2$C $\longrightarrow$ B^{n-1} / O / —OB / —O—P(=O)(O)—O—CH$_2$ / C.OCH$_2$; B^n / O / —OD / —OD $\xrightarrow{\text{selective removal of C, etc.}}$

Care must be taken, however, to ensure that cyclic phosphate formation, with possible acyl transfer from the 3-hydroxyl to the 2-hydroxyl position, does not occur.

All of the protecting groups **A–C** are of the hydroxyl protecting type. Various acyl-protecting groups, for example HCO– and $CMe_3.CO$–, the trityl group, the tetrahydropyranyl group and, more recently, the 4-methoxytetrahydropyranyl group, have proved useful. Typically, base-labile 5-hydroxyl and acid-labile 3-hydroxyl groups (or *vice versa*) have been employed.

The 3-hydroxyl terminal nucleoside can be blocked by an alkylidene group (**DD**) since selective removal of the 3-hydroxyl protecting group is not required in this situation: the methoxymethylidene group

O O
H OMe

has been used in this way.

Reactive functions in the base components of the nucleosides can be blocked by the standard protecting groups used in peptide synthesis.

N,N'-Dicyclohexylcarbodiimide has been used in the formation of the phosphate ester links, as have mixed anhydrides prepared from aryl sulphonyl chlorides in pyridine solution. The danger here is that dimerization will occur and monophosphorylating agents are therefore to be preferred. One such reagent is phenylphosphoryldichloridate, $PhO.POCl_2$, from which a first chlorine atom is displaced more readily than the second; the phenyl ester is cleaved under alkaline conditions at the end of the synthesis.

Solid-phase methods have been tried in nucleotide synthesis, but, so far, they have not been as strikingly successful as in peptide synthesis.

Index

NOTE: Common amino acids are arranged under their initial letter, e.g. *isoleucine* is found under *I*; less common amino acids are arranged as derivatives, e.g. *hydroxylysine* is to be found under *lysine*. Protecting groups and activating groups are arranged under the relevant functional group, e.g. *benzyloxycarbonyl* is to be found under *amino group protection.*

particularly labile, other conditions of hydrolysis can be important. Alkaline hydrolysis is sometimes used and, for example, in the case of peptides containing tryptophan, it can be essential. The chances of obtaining intact acyl serine peptide bonds are clearly better under alkaline conditions. On the other hand, extensive racemization of amino acids might be expected if the conditions are vigorous and, in some instances, base-catalysed eliminations from amino acid side chains (e.g. **30** → **31**) lead to further complications.

$$\underset{\mathbf{(30)}}{R^1.NH.\overset{\overset{CH_2OH}{|}}{C}H.CO.NH.R^2} \longrightarrow \underset{\mathbf{(31)}}{R^1.NH.\overset{\overset{CH_2}{\|}}{C}.CO.NH.R^2}$$

Living organisms utilize enzymes called peptidases to degrade peptides. Fundamentally, these enzymes are of two different types: exopeptidases cleave only terminal peptide bonds, whilst endopeptidases cleave only internal peptide bonds. It will be seen shortly that some of the first class of enzymes find application in terminal residue and terminal sequence determination; members of the second group play an important complementary role to partial acid hydrolysis in the fragmentation of the parent peptide. In addition to the broad specificities implicit in the classification of the peptidases, individual enzymes generally have their own particular specific requirements concerning the bonds they will hydrolyse. For example, trypsin hydrolyses peptide bonds which involve the carboxyl group of basic amino acid residues; chymotrypsin, peptide bonds which involve the carboxyl group of aromatic amino acid residues. These specificities are not rigid and unexpected cleavage at other residues sometimes occurs. Many enzymes, for example, pronase, nagarse and papaine, are even less rigorous in their requirements but, usually, some particular peptide bonds are cleaved so much more rapidly than others by the action of endopeptidases that a fragmentation of the parent peptide into a number of oligopeptides can be achieved.

Terminal residue identification

*Exercise 2.4**

Attempt to devise chemical procedures with which it might prove possible to identify either *N* or *C*-terminal residues in peptides.

Of the several ways in which terminal residues might be identified, the most obvious involves modifying the terminal residue so that, after total hydrolysis of the peptide, it can be distinguished from the unmodified amino acids produced. Selective modification of the terminal residue will be dependent on the unique possession by this residue of some function or property not possessed by other amino acid residues in the peptide. The α-amino group is a distinctive feature of the *N*-terminal residue, the α-carboxyl group of the *C*-terminal residue. Any reaction which an α-amino or α-carboxyl group will undergo selectively in the presence of the peptide bonds can thus be considered for terminal residue identification. It is imperative that the modification of the terminal residue proceeds without cleavage of peptide bonds because, otherwise, artefact terminal residues would be produced.

Often, only small amounts of the original peptide or protein are available for structural studies and this means that the peptides produced by hydrolysis will be isolated on an even smaller scale. Terminal residue identification must therefore be a microprocedure and the best techniques which are available can be applied to minute amounts of material. Obvious ways in which a very sensitive identification of the terminal residue might be achieved would involve converting it either to a highly coloured or to a radioactive derivative.

The most widely used technique for the identification of *N*-terminal residues employs 2,4-dinitrofluorobenzene (**32**) to form the 2,4-dinitrophenyl (DNP*) peptide (**33**). Subsequent hydrolysis yields the *N*-terminal residue as its DNP-derivative (**34**).

$$2,4\text{-}(NO_2)_2C_6H_3F + NH_2.CHR^1.CO.NH.CHR^2.CO\ldots NH.CHR^n.CO_2^- \longrightarrow$$

(**32**)

$$2,4\text{-}(NO_2)_2C_6H_3NH.CHR^1.CO.NH.CHR^2.CO\ldots NH.CHR^n.CO_2^- \xrightarrow[H_2O]{H^+}$$

(**33**)

$$2,4\text{-}(NO_2)_2C_6H_3NH.CHR^1.CO_2H + \overset{+}{N}H_3.CHR^2.CO_2H\ldots + \overset{+}{N}H_3.CHR^n.CO_2H$$

(**34**)

* Not to be confused with dinitrophenylhydrazone derivatives commonly abbreviated in the same way